T0075327

SpringerBriefs in Mathematics

Series Editors

Palle Jorgensen, Iowa, USA
Roderick Melnik, Waterloo, Canada
Lothar Reichel, Kent, USA
George Yin, Detroit, USA
Nicola Bellomo, Torino, Italy
Michele Benzi, Pisa, Italy
Tatsien Li, Shanghai, China
Otmar Scherzer, Linz, Austria
Benjamin Steinberg, New York, USA
Yuri Tschinkel, New York, USA
Ping Zhang, Kalamazoo, USA

SpringerBriefs in Mathematics showcases expositions in all areas of mathematics and applied mathematics. Manuscripts presenting new results or a single new result in a classical field, new field, or an emerging topic, applications, or bridges between new results and already published works, are encouraged. The series is intended for mathematicians and applied mathematicians.

More information about this series at http://www.springer.com/series/10030

Shubham Dwivedi · Jonathan Herman ·
Lisa C. Jeffrey · Theo van den Hurk

Hamiltonian Group Actions and Equivariant Cohomology

 Springer

Shubham Dwivedi
Department of Pure Mathematics
University of Waterloo
Waterloo, ON, Canada

Jonathan Herman
Department of Mathematical and
Computational Sciences
University of Toronto at Mississauga
Mississauga, ON, Canada

Lisa C. Jeffrey
Department of Mathematics
University of Toronto
Toronto, ON, Canada

Theo van den Hurk
Department of Mathematics
University of Toronto
Toronto, ON, Canada

ISSN 2191-8198   ISSN 2191-8201   (electronic)
SpringerBriefs in Mathematics
ISBN 978-3-030-27226-5   ISBN 978-3-030-27227-2   (eBook)
https://doi.org/10.1007/978-3-030-27227-2

Mathematics Subject Classification (2010): 57R17

© The Author(s), under exclusive licence to Springer Nature Switzerland AG 2019
This work is subject to copyright. All rights are solely and exclusively licensed by the Publisher, whether the whole or part of the material is concerned, specifically the rights of translation, reprinting, reuse of illustrations, recitation, broadcasting, reproduction on microfilms or in any other physical way, and transmission or information storage and retrieval, electronic adaptation, computer software, or by similar or dissimilar methodology now known or hereafter developed.
The use of general descriptive names, registered names, trademarks, service marks, etc. in this publication does not imply, even in the absence of a specific statement, that such names are exempt from the relevant protective laws and regulations and therefore free for general use.
The publisher, the authors and the editors are safe to assume that the advice and information in this book are believed to be true and accurate at the date of publication. Neither the publisher nor the authors or the editors give a warranty, expressed or implied, with respect to the material contained herein or for any errors or omissions that may have been made. The publisher remains neutral with regard to jurisdictional claims in published maps and institutional affiliations.

This Springer imprint is published by the registered company Springer Nature Switzerland AG
The registered company address is: Gewerbestrasse 11, 6330 Cham, Switzerland

Preface

This monograph could serve as the textbook for a graduate course on symplectic geometry. It has been used for this purpose in graduate courses taught in the Mathematics Department at the University of Toronto. This book evolved from the lecture notes for the graduate course on symplectic geometry taught in fall 2016. Three of the coauthors (Shubham Dwivedi, Jonathan Herman and Theo van den Hurk) were students in this course.

Alternatively, it could be used for independent study. A course on differential topology is an essential prerequisite for this course (at the level of the texts by Boothby [1] or Lee [2]). Some of the later material will be more accessible to readers who have had a basic course on algebraic topology, at the level of the book by Hatcher [3]. For some of the later chapters (such as the chapter on geometric quantization and the chapter on flat connections on 2-manifolds), it would be helpful to have some background on representation theory (such as the book by Bröcker and tom Dieck [4]) and complex geometry (such as the first chapter of the book by Griffiths and Harris [5]).

The layout of this monograph is as follows. The first chapter introduces symplectic vector spaces, followed by symplectic manifolds. The second chapter treats Hamiltonian group actions. The Darboux theorem comes in Chap. 3. Chapter 4 treats moment maps. Orbits of the coadjoint action are introduced in Chap. 5.

Chapter 6 treats symplectic quotients.

The convexity theorem (Chap. 7) and toric manifolds (Chap. 8) come next. Equivariant cohomology is introduced in Chap. 9.

The Duistermaat–Heckman theorem follows in Chap. 10, geometric quantization in Chap. 11 and flat connections on 2-manifolds in Chap. 12. Finally, the appendix provides the background material on Lie groups.

Exercises related to the material presented here may be found at

Chaps. 1–5

http://www.math.toronto.edu/~jeffrey/mat1312/exerc1.pdf

Chaps. 6–8, 11

http://www.math.toronto.edu/~jeffrey/mat1312/exerc2rev.pdf

Chaps. 9, 10, 12

http://www.math.toronto.edu/~jeffrey/mat1312/exerc3rev.pdf

The authors thank Yucong Jiang and Caleb Jonker for carefully reading and commenting on several chapters of the manuscript.

Waterloo, Canada   Shubham Dwivedi
Mississauga, Canada   Jonathan Herman
Toronto, Canada   Lisa C. Jeffrey
Toronto, Canada   Theo van den Hurk

References

1. W. Boothby, *An Introduction to Differentiable Manifolds and Riemannian Geometry*. Pure and Applied Mathematics, vol. 120 (Academic Press, New York, 1986)
2. J. Lee, *Introduction to Smooth Manifolds*. GTM (Springer, New York, 2006)
3. A. Hatcher, *Algebraic Topology* (Cambridge University Press, Cambridge, 2001)
4. T. Bröcker, T. tom Dieck, *Representations of Compact Lie Groups*. GTM (Springer, New York, 1985)
5. P. Griffiths, J. Harris, *Principles of Algebraic Geometry* (Wiley, New Jersey, 1994)

Contents

Notations

ω	Symplectic form
d	Exterior derivative
L_X	Lie derivative with respect to X
J	Almost complex structure
$\Gamma(TM)$	Sections of tangent bundle
$\Gamma(E)$	Sections of the bundle E
i_X	Interior product with respect to X
Φ	Moment map
μ	Moment map
$C^\infty(M)$	Smooth real-valued functions on M
D	Differential in the Cartan model

Chapter 1
Symplectic Vector Spaces

This chapter is a brief introduction to symplectic manifolds. We will start this chapter by defining a symplectic vector space (Sect. 1.1). After briefly reviewing the notion of an almost complex structure on a vector space, we will see how the compatibility condition between the symplectic form and an almost complex structure gives rise to an inner product. In Sect. 1.3, we will discuss the definition of symplectic manifolds, describe some of their basic properties and will finally see some examples in Sect. 1.4. Section 1.2 contains a review of results from differential topology which are essential material for what follows.

1.1 Properties of Symplectic Vector Spaces

Definition 1.1 Let V be a n-dimensional vector space over $\mathbb{R}$ and let $\omega : V \times V \to \mathbb{R}$ be a bilinear map. The map ω is called **skew-symmetric** if $\omega(u, v) = -\omega(v, u)$, for all $u, v \in V$.

From a bilinear form ω on V, we get a linear map $\tilde{\omega} : V \to V^*$, where V^* is the dual space to V, by

$$\tilde{\omega}(u)(v) = \omega(u, v) \quad \text{for all } u, v \in V$$

Definition 1.2 A bilinear map ω is called **non-degenerate** if $\tilde{\omega}$ is a bijection.

We have the following theorem about skew-symmetric bilinear maps, whose proof can be found in [1]

Theorem 1.3 *Let ω be a skew-symmetric bilinear map on V. There exists a basis $\{u_1, ..., u_k, e_1, ..., e_m, f_1, ..., f_m\}$ of V such that*

© The Author(s), under exclusive licence to Springer Nature Switzerland AG 2019
S. Dwivedi et al., *Hamiltonian Group Actions and Equivariant Cohomology*,
SpringerBriefs in Mathematics,
https://doi.org/10.1007/978-3-030-27227-2_1

$$\omega(u_i, v) = 0 \quad for\ all\ i\ and\ all\ v \in V$$
$$\omega(e_i, e_j) = 0 \quad for\ all\ i, j$$
$$\omega(f_i, f_j) = 0 \quad for\ all\ i, j$$
$$\omega(e_i, f_j) = \delta_{ij} \quad for\ all\ i, j$$

where δ is the Kronecker delta.

We can now define a **symplectic form** on a vector space.

Definition 1.4 A skew-symmetric and non-degenerate bilinear form $\omega :$ $V \times V \to \mathbb{R}$ is called a symplectic form on V and (V, ω) is called a symplectic vector space.

Note that since ω is skew-symmetric and non-degenerate, $\tilde{\omega}$ must be a bijection and hence

$$\ker \tilde{\omega} = \{u \in V \mid \omega(u, v) = 0, \ \text{for all } v \in V\}$$

must be $\{0\}$ and thus k in Theorem 1.3 must be zero. Thus we see from Theorem 1.3 that V must be even-dimensional. A basis of the form in Theorem 1.3 is called a *symplectic basis*.

Remark 1.5 We may form $\omega^n : \Lambda^{2n} V \to \mathbb{R}$. Now $\Lambda^{2n} V$ is a one-dimensional vector space, so ω^n uniquely determines a real number (a multiple of the top exterior power of the elements of a symplectic basis).

Let us now make some definitions which will be used later.
Let (V, ω) be a symplectic vector space.

Definition 1.6 An **almost complex structure** $J : V \to V$ is a linear map such that $J^2 = -I$.

This is what we mean by multiplication by i; in other words, we can make a $\mathbb{R}$ vector space into a $\mathbb{C}$ vector space by defining $(a + ib)x = ax + bJ(x)$ for $a, b \in \mathbb{R}$, $x \in V$. We can use this to identify $V \cong \mathbb{R}^{2n}$ with $\mathbb{C}^n$.

Definition 1.7 An almost complex structure on V is said to be *compatible* with the symplectic structure if

$$\omega(JX, JY) = \omega(X, Y)$$

for all $X, Y \in V$.

The compatibility condition on the almost complex structure J allows us to define a symmetric bilinear form $< \cdot, \cdot >$ on the vector space. However, this symmetric bilinear form is only an inner product if the almost complex structure satisfies an additional condition, positivity.

Definition 1.8 An almost complex structure J which is compatible with the symplectic form ω is *positive* if for all $X \neq 0$,

$$\omega(X, JX) > 0.$$

The reason for calling this positive is evident from the following proposition:

Proposition 1.9 *A symplectic form ω on a vector space V together with a compatible positive almost complex structure J determines an inner product*

$$< \cdot, \cdot >$$

on V, by
$$< X, Y >= \omega(X, JY).$$

Proof Since J is a compatible almost complex structure, we have

$$\omega(Y, JX) = -\omega(JX, Y) = \omega(JX, J^2Y) = \omega(X, JY)$$

which gives that $< \cdot, \cdot >$ is symmetric. The positive definiteness of $< \cdot, \cdot >$ follows from the fact that J is positive. $\qquad\qquad\qquad\qquad\qquad\qquad\qquad\square$

Let us see the prototype example of a symplectic vector space.

Example 1.10 Let
$$V = \mathbb{R}^{2n}.$$

Then V has a symplectic basis $\{e_1, \ldots, e_n, f_1, \ldots, f_n\}$; in other words, if ω is the symplectic form, then we have $\omega(e_i, e_j) = 0$, $\omega(f_i, f_j) = 0$ and $\omega(e_i, f_j) = \delta_{ij}$, for all i, j. A suitable almost complex structure compatible with this symplectic form is

$$Je_i = f_i, \quad Jf_i = -e_i$$

or

$$J = \begin{bmatrix} 0 & -I \\ I & 0 \end{bmatrix}.$$

The compatible inner product has $\{e_1, \ldots, e_n, f_1, \ldots, f_n\}$ as an orthonormal basis.
    The group of transformations preserving the symplectic form is

$$\mathrm{Sp}(V) = \left\{ \begin{bmatrix} A & B \\ C & D \end{bmatrix} \in \mathrm{GL}(2n, \mathbb{R}) \,|\, A^T C, \; B^T D \text{ are symmetric}, \; A^T D - C^T B = I \right\}$$

where A^T denotes the transpose of A.
    To see this, note that the condition for $R \in \mathrm{Sp}(2n, \mathbb{R})$ is

$$\omega(RX, RY) = \omega(X, Y) \quad \text{for all } X, Y$$

which is equivalent to

$$\langle JRX, RY \rangle = \langle JX, Y \rangle$$

or

$$\langle R^t JRX, Y \rangle = \langle JX, Y \rangle$$

in other words

$$R^T JR = J$$

or

$$\begin{bmatrix} A^T & C^T \\ B^T & D^T \end{bmatrix} \begin{bmatrix} 0 & -I \\ I & 0 \end{bmatrix} \begin{bmatrix} A & B \\ C & D \end{bmatrix} = \begin{bmatrix} 0 & -I \\ I & 0 \end{bmatrix}.$$

The subgroup

$$\left\{ \begin{bmatrix} A & -B \\ B & A \end{bmatrix} \mid A^*A + B^*B = I, \ B^T A = A^T B \right\}$$

where A^* is the conjugate transpose of A is isomorphic to $U(n)$ under

$$(A, B) \mapsto A + iB$$

This subgroup consists of all linear transformations preserving ω and commuting with $\begin{bmatrix} 0 & -I \\ I & 0 \end{bmatrix}$.

1.2 Review of Results From Differential Topology

We will assume familiarity with the following concepts. Excellent sources for these materials are [2] and [3]

1. differential manifolds,
2. tangent spaces $T_m M$,
3. tangent bundles TM and cotangent bundles T^*M,
4. line bundles,
5. vector fields (sections of the tangent bundle), and
6. differential forms (sections of exterior powers of the cotangent bundle),

    a. wedge product,
    b. exterior derivative d, $d \circ d = 0$,
    c. interior product i_X with vector field X, and

d. Lie derivative with respect to a vector field and Cartan's formula

$$L_X = di_X + i_X d$$

1.3 Symplectic Manifolds

Let M^n be a n-dimensional manifold and ω be a 2-form on M, that is, for every $p \in M$, $\omega_p : T_pM \times T_pM \to \mathbb{R}$ is a skew-symmetric bilinear map from the tangent space to M at the point p to $\mathbb{R}$, and ω_p varies smoothly in p.

We say that a 2-form ω is closed if $d\omega = 0$ where d is the exterior derivative on M.

Definition 1.11 A 2-form ω on M is called **symplectic** if $d\omega = 0$ and ω_p is a symplectic for every $p \in M$.

Definition 1.12 A **symplectic manifold** is a manifold M equipped with a symplectic form ω.

From the discussion following Definition 1.14, we see that a symplectic manifold is even-dimensional.

We defined an almost complex structure on a vector space V. In the same way, we can give the following definition.

Definition 1.13 An *almost complex structure* on M is a section J of $\text{End}(TM)$ such that $J^2 = -I$, which is to say that for every $p \in M$, $J_p : T_pM \to T_pM$ is a linear map with $J_p{}^2 = -I$ on T_pM and J_p varies smoothly in p.

Definition 1.14 The *symplectic volume* on M is the form $\omega^n/n!$ of top dimension n.

Since the symplectic form is non-degenerate, $\dfrac{\omega^n}{n!}$ is a nowhere vanishing form on M and its existence means that symplectic manifolds are oriented.

Remark 1.15 An almost complex structure J on a symplectic manifold M is *integrable* if it comes from a structure of complex manifold on M (in other words the transition functions between charts on M are holomorphic functions). The almost complex structure J represents multiplication by i on the holomorphic tangent space.

See for example Chap. 0 of Griffiths–Harris [4].

Definition 1.16 If (M, ω) is a symplectic manifold with a compatible positive almost complex structure J and the almost complex structure is a complex structure (i.e. integrable), then M is called a Kähler manifold.

Suppose (M, ω) is a symplectic manifold with an almost complex structure J (a section of $\text{End}(TM)$ satisfying $J^2 = -\text{Id}$). Then, a Riemannian metric on M is obtained by

$$g(X, Y) = \omega(X, JY)$$

for all $X, Y \in \Gamma(TM)$ (here the symbol $\Gamma(TM)$ means a section of the tangent bundle, i.e., a vector field). In fact, any two of $\{\omega, J, g\}$ which are compatible uniquely determine the third.

$$\omega(X, Y) = \omega(JX, JY) = g(JX, Y)$$

and so on.

1.4 Examples

Let us see some examples of symplectic manifolds.

1. Let $M = \mathbb{R}^{2n}$ with coordinates $x_1, \dots, x_n, y_1, \dots, y_n$. The 2-form

$$\omega_0 = \sum_{i=1}^{n} dx_i \wedge dy_i$$

is a symplectic form. Thus $(\mathbb{R}^{2n}, \omega_0)$ is a symplectic manifold. In fact, we will later see (Darboux's theorem, in Chap. 3) that any $2n$-dimensional symplectic manifold locally looks like $(\mathbb{R}^{2n}, \omega_0)$.

2. Consider $\mathbb{C}^n$ with coordinates $z_1, \dots, z_n$. The 2-form

$$\omega = \frac{\sqrt{-1}}{2} \sum_{i=0}^{n} dz_i \wedge d\bar{z}_i$$

is a symplectic form.

3. Consider the 2-sphere S^2 as a subset of unit vectors in $\mathbb{R}^3$. For any point $s \in S^2$, $T_s S^2$ is the set of vectors which are orthogonal to the unit vector s. The standard symplectic form on S^2 is given as

$$\omega_s(u, v) = \langle s, u \times v \rangle \qquad \text{for all } u, v \in T_s S^2$$

where $\times$ is the cross product on $\mathbb{R}^3$. Since ω is a 2-form and S^2 is two-dimensional, it follows that ω is closed.

References

1. A.C. Da Silva, *Lectures on Symplectic Geometry*, vol. 1764, Lecture Notes in Mathematics (Springer, Berlin, 2001)
2. W. Boothby, *An Introduction to Differentiable Manifolds and Riemannian Geometry*, vol. 120, Pure and Applied Mathematics (Academic Press, Cambridge, 1986)
3. J. Lee, *Introduction to Smooth Manifolds*, GTM (Springer, Berlin, 2006)
4. P. Griffiths, J. Harris, *Principles of Algebraic Geometry* (Wiley, Hoboken, 1994)

References

Chapter 2
Hamiltonian Group Actions

In this chapter, we will define Hamiltonian flows, Hamiltonian actions and moment maps.

The layout of the chapter is as follows. In Sect. 2.1 we recall the original example of a Hamiltonian flow, namely, Hamilton's equations. In Sect. 2.2, we will start by understanding what Hamiltonian vector fields and Hamiltonian functions are. In Sect. 2.3, we will introduce a bracket on the set of smooth functions on a symplectic manifold which will satisfy the Jacobi identity and will make the former into a Lie algebra. We will see some examples of such vector fields on S^2 and the 2-torus. In the final section (Sect. 2.4), we will define a moment map and will list some conditions which will guarantee the existence of moment maps and other conditions which guarantee their uniqueness.

2.1 Hamilton's Equations

The basic example of a Hamiltonian flow gives rise to Hamilton's equations. Consider $\mathbb{R}^2$ with coordinates (p, q). In physics, q denotes the position of an object and p denotes the momentum (in other words $p = mdq/dt$, where t denotes time and m is the mass of the object).

Let the Hamiltonian H be defined by

$$H(p, q) = \frac{p^2}{2m} + V(q)$$

where $\frac{p^2}{2m}$ is the kinetic energy and $V(q)$ is the potential energy (which depends only on q, not on p). The Hamiltonian is the sum of kinetic and potential energy. The symplectic form on $\mathbb{R}^2$ is

$$\omega = dq \wedge dp.$$

© The Author(s), under exclusive licence to Springer Nature Switzerland AG 2019
S. Dwivedi et al., *Hamiltonian Group Actions and Equivariant Cohomology*,
SpringerBriefs in Mathematics,
https://doi.org/10.1007/978-3-030-27227-2_2

Then, the Hamiltonian vector field X_H is defined by the property that its contraction with the symplectic form equals the differential of H:

$$i_{X_H}\omega = dH$$

which tells us that

$$X_H = H_p \frac{\partial}{\partial q} - H_q \frac{\partial}{\partial p}.$$

If $(q(t), p(t))$ is a path in $\mathbb{R}^2$ which integrates this vector field, we find that it obeys the following equations:

$$\frac{dq}{dt} = H_p = \frac{p}{m} \tag{2.1}$$

$$\frac{dp}{dt} = -H_q = -\frac{dV}{dq} \tag{2.2}$$

The first equation above tells us only the definition of momentum (as the product of mass times the velocity $\frac{dq}{dt}$). The second equation is Newton's law of motion (that the force on the object, namely, $-\frac{dV}{dt}$, is equal to the first derivative of the momentum; in other words, the mass times the acceleration).

2.2 Hamiltonian Flow of a Function

Let (M, ω) be a symplectic manifold, and $H : M \to \mathbb{R}$ a C^∞ function. Then $dH \in \Omega^1(M)$. Recall that if we have a symplectic form on a vector space V then it induces a bijection between V and its dual V^*. The symplectic form ω on M is non-degenerate, so for all $m \in M$ there exists a unique vector field X_H such that

$$(dH)_m = i_{(X_H)_m}\omega_m$$

and hence

$$dH = i_{X_H}\omega \tag{2.3}$$

We have the following.

Definition 2.1 A vector field X_H as above is called the **Hamiltonian vector field** and H is called its **Hamiltonian function**.

Conversely, if we are given a vector field ξ on M, ω determines a 1-form α on M by $i_\xi\omega = \alpha$.

Suppose the Lie derivative of ω with respect to the vector field ζ is 0. (This means the transformation $M \to M$ obtained by integrating the vector field preserves ω.) Then Cartan's formula tells us that

$$0 = L_\zeta \omega = d i_\zeta \omega + i_\zeta d\omega$$

and the last term is zero because ω is closed. So $i_\zeta \omega$ is closed. We shall see that $i_\zeta \omega$ is *exact* if and only if the vector field ζ arises as the Hamiltonian vector field of a smooth function $H : M \to \mathbb{R}$. We get the following proposition from (2.3).

Proposition 2.2 *Suppose (M, ω) is a symplectic manifold and X_H is a Hamiltonian vector field with Hamiltonian function H. Then for $m \in M$,*

$$dH_m = 0$$

(m is a critical point of H) if and only if $(X_H)_m = 0$.

2.3 Poisson Bracket

We are going to define the *Poisson Bracket* of two functions. A good reference for this section is the book [1, Chap. 2] by Audin.

Suppose

$$f, g : M \to \mathbb{R}$$

are smooth functions. We define

$$\{f, g\} := \omega(X_f, X_g)$$

where X_f and X_g are Hamiltonian vector fields corresponding to f and g, respectively. The pairing $\{f, g\}$ is called the *Poisson bracket* of f, g. The reason for calling this a "bracket" is evident in the following theorem.

Theorem 2.3 *Let $f, g \in C^\infty(M)$. Then*

$$X_{\{f,g\}} = -[X_f, X_g]$$

where $[\cdot, \cdot]$ is the Lie bracket of vector fields.

Proof For $X, Y \in \Gamma(TM)$

$$i_{[X,Y]}\omega = L_X i_Y \omega - i_Y L_X \omega$$
$$= (d i_X + i_X d) i_Y \omega - i_Y (d i_X + i_X d)\omega$$

The first line in the above equation follows from [2, V.8, Exercise 8], which reduces (in the case of a 2-form ω and vector fields X, U, V) to

$$(L_X\omega)(U, V) = L_X\left(\omega(U, V)\right) - \omega(L_X U, V) - \omega(U, L_X V).$$

The second line follows from Cartan's formula for the Lie derivative.

However. if $X = X_f, Y = X_g$ are Hamiltonian vector fields for f and g, then $i_{X_f}\omega = df, i_{X_g}\omega = dg$ so using $d \circ d = 0$ we have

$$\begin{aligned} i_{[X_f, X_g]}\omega &= d(i_X i_Y \omega) - d(i_Y i_X \omega) \\ &= -d(\omega(X_f, X_g)) \\ &= -d\{f, g\}. \end{aligned} \qquad \square$$

Proposition 2.4 *The Poisson bracket satisfies the Jacobi identity, i.e.*

$$\{f_1, \{f_2, f_3\}\} + \{f_2, \{f_3, f_1\}\} + \{f_3, \{f_1, f_2\}\} = 0$$

for all $f_1, f_2, f_3 \in C^\infty(M)$.

Proof If X_i denotes the Hamiltonian vector field for f_i, then

$$\begin{aligned} 0 &= d\omega(X_1, X_2, X_3) \\ &= X_1\omega(X_2, X_3) - X_2\omega(X_1, X_3) + X_3\omega(X_1, X_2) - \omega([X_1, X_2], X_3) \\ &\quad + \omega([X_1, X_3], X_2) - \omega([X_2, X_3], X_1) \end{aligned}$$

Now $X_1\omega(X_2, X_3) = X_1 \cdot \{f_2, f_3\}$ and $X_1 \cdot \{f_2, f_3\} = \{\{f_2, f_3\}, f_1\}$. The last identity follows because the Poisson bracket is defined by

$$\{f_1, f_2\} = \omega(X_1, X_2)$$

where X_1 (resp. X_2) is the Hamiltonian vector field of the function f_1 (resp. f_2). The proof then follows. $\square$

From Theorem 2.3 and Proposition 2.4, we get the following.

Proposition 2.5 $C^\infty(M)$ *is a Lie algebra under $\{\cdot, \cdot\}$ and the map $f \mapsto X_f$ is a Lie algebra homomorphism.*

Proof The only nontrivial task is to check that the map takes Poisson bracket of functions to Lie brackets of their images, but that is obvious from Theorem 2.3 and Proposition 2.4. $\square$

Vector Fields Arising From Group Actions:
We study two vector fields on two-dimensional manifolds whose flows are area-preserving.

Example 2.6 The group $U(1)$ acts on S^2 by rotation around the vertical axis.

Coordinates on $S^2 \setminus \{N, S\}$ are (z, ϕ) where z is the height function and ϕ is the angle relative to the x-axis (normally called the azimuthal angle) and the symplectic form is

$$\omega = dz \wedge d\phi.$$

We restrict to $z \in (-1, 1)$ and $\phi \in [0, 2\pi)$.
The vector field generated by the $U(1)$ action is $\frac{\partial}{\partial \phi}$ and

$$i_{\frac{\partial}{\partial \phi}} \omega = -dz.$$

In this case, $H = z : S^2 \to \mathbb{R}$ is a smooth function and $X_H = \frac{\partial}{\partial \phi}$. This is the interior product of the symplectic form with $\frac{\partial}{\partial \phi}$ so this is a Hamiltonian action (see Definition 2.9).

Example 2.7 The group $U(1)$ acts on the 2-torus $\{(e^{i\theta}, e^{i\phi})\}$ by rotation in the ϕ direction. Again $\omega = d\theta \wedge d\phi$.

The 1-forms generated by these vector fields are as follows. Note that

$$i_{\frac{\partial}{\partial \phi}} \omega = -d\theta.$$

But $\theta : S^1 \times S^1 \to [0, 2\pi]$ is not a continuous $\mathbb{R}$-valued function (rather a $\mathbb{R}/2\pi\mathbb{Z}$-valued function). The form $d\theta$ is closed but not exact. In fact

$$H^1(S^1 \times S^1) = \mathbb{R} \oplus \mathbb{R}$$

and the de Rham cohomology class of the form $[-d\theta]$ is one of the generators (the other is $[d\phi]$). So in this case the vector field does not come from the Hamiltonian flow of a smooth real-valued function.

Example 2.8 Introduce polar coordinates (r, ϕ) on $\mathbb{R}^2 \setminus \{0\}$. The vector field $\frac{\partial}{\partial \phi}$ preserves the symplectic form. The vector field $\frac{\partial}{\partial r}$ does not preserve the symplectic form. In these coordinates the symplectic form is

$$\omega = r dr \wedge d\phi = \frac{1}{2} d(r^2) \wedge d\phi.$$

Suppose G is a Lie group. Recall that a Lie group action on a manifold M means a smooth map $F : G \times M \to M$ (define also the diffeomorphisms $F_g : M \to M$ by $F_g(m) = F(g, m)$) so that the maps F_g satisfy

$$F_{gh} = F_g \circ F_h$$

and that the identity element e of G acts by the identity map, and the inverse g^{-1} acts by $F_{g^{-1}} = (F_g)^{-1}$. Examples include the rotation action of the circle group $U(1)$ on the complex plane (by multiplication). For more information about Lie groups, see Appendix.

Definition 2.9 Let G be a Lie group with Lie algebra $\mathbf{g}$. Suppose G acts on a symplectic manifold M and that the action preserves the symplectic form. This means $L_{\tilde{X}}\omega = 0$ where $\tilde{X}$ is the vector field associated to the action of any $X \in \mathbf{g}$.

The action is called *Hamiltonian* if there is a collection of smooth functions $\Phi_\eta : M \to \mathbb{R}$ for all $\eta \in \mathbf{g}$ satisfying

1. $X_{\Phi_\eta} = \eta^\sharp$ (in other words the Hamiltonian vector field associated to Φ_η is the vector field $\eta^\sharp$ generated by the action of $\eta \in \mathbf{g}$).
2. The Φ_η are derived from a function $\Phi : M \to \mathbf{g}^*$ via $\Phi_\eta(m) = \Phi(m)(\eta)$ for all $m \in M, \eta \in \mathbf{g}$.
3. The map Φ is G-equivariant, where G acts on $\mathbf{g}^*$ via the coadjoint action.[1]

Remark 2.10 If G is abelian, then Φ being equivariant just means that Φ is invariant under the G action.

Remark 2.11 The map Φ defined above is called a **moment map** for the action of G on M.

Remark 2.12 We shall consider only actions of compact connected Lie groups, although the definition of Hamiltonian actions may be extended to actions by non-compact groups. In particular, the term "torus" refers to the compact connected abelian group $T \cong U(1)^n$.

2.4 Uniqueness

The moment map is defined as

$$d\Phi^X = i_{X^\sharp}\omega.$$

In other words, Φ^X is the Hamiltonian function generating the vector field $X^\sharp$.

We also require that $\Phi : M \to \mathbf{g}^*$ is equivariant (where G acts on $\mathbf{g}^*$ by the coadjoint action).

As we have seen, moment maps for group actions preserving the symplectic structure do not always exist. Some conditions which guarantee their existence include

---

[1] If $\langle , \rangle : \mathbf{g}^* \times \mathbf{g} \to \mathbb{R}$ is the natural pairing then for any $\psi \in \mathbf{g}^*$ we define $Ad_g{}^*\psi$ by $\langle Ad_g{}^*\psi, X \rangle = \langle \psi, Ad_{g^{-1}}X \rangle$ for any $X \in \mathbf{g}$. Hence, we get a map $Ad^* : G \to \mathrm{GL}(\mathbf{g}^*)$, known as the coadjoint action of G on $\mathbf{g}^*$.

1. M is simply connected (because $H^1(M, \mathbb{R}) = 0$, closed 1-forms are necessarily exact).
2. Lie(G) is *semisimple* (for example $G = SU(n)$).

If G is abelian, the moment map is never unique, because it is always possible to replace the moment map by an equivalent moment map by adding a constant to it, and the new moment map is still equivariant. On the other hand, if G is semisimple, adding a constant to the moment map produces a map that is no longer equivariant, so in this case the moment map is unique.

For more details on conditions guaranteeing the existence of moment maps, and other conditions guaranteeing uniqueness of moment maps, see [3, Chaps. 24–26].

We denote the set of Hamiltonian vector fields by $\mathcal{H}$. There is an exact sequence

$$0 \to \mathbb{R} \to C^\infty(M) \to \mathcal{H} \to 0.$$

Here, the third arrow sends a function f to X_f, the Hamiltonian vector field associated to f. A second exact sequence is

$$0 \to \mathcal{H} \to S \to H^1(M, \mathbb{R}) \to 0.$$

Here, S denote the set of symplectic vector fields (in other words $X \in \Gamma(TM)$ is in S if $L_X \omega = 0$). They are associated with closed 1-forms, and Hamiltonian vector fields are associated with exact 1-forms; hence, the quotient of the two is the first de Rham cohomology group of M.

References

1. M. Audin, *Torus Actions on Symplectic Manifolds*, vol. 93, Progress in Mathematics (Birkhäuser, Basel, 2004)
2. W. Boothby, *An Introduction to Differentiable Manifolds and Riemannian Geometry*, vol. 120, Pure and Applied Mathematics (Academic Press, Cambridge, 1986)
3. V. Guillemin, S. Sternberg, *Symplectic Techniques in Physics* (Cambridge University Press, Cambridge, 1986)

Chapter 3
The Darboux–Weinstein Theorem

Informally, the Darboux–Weinstein theorem says that given any two symplectic manifolds of the same finite dimension, they look alike locally. It states that around any point of a symplectic manifold, there is a chart for which the symplectic form has a particularly nice form. In this section, we give a proof of an equivariant version of the theorem and look at some corollaries. We direct the reader to [1] or Sect. 22 of [2] for more details.

Theorem 3.1 (Darboux–Weinstein) *Suppose ω is a symplectic form on a manifold M^{2n}. Then for any $x \in M$ there is a neighbourhood U of x and a diffeomorphism $\phi : U \to \mathbb{R}^{2n}$ such that*

$$\phi^* \left(\sum_i dq_i \wedge dp_i \right) = \omega,$$

where $q_1, \cdots, q_n, p_1, \cdots, p_n$ are the standard coordinates on $\mathbb{R}^{2n}$.

An equivariant version of the Darboux–Weinstein theorem is as follows.

Theorem 3.2 (Equivariant Darboux–Weinstein) *Suppose $N \subset M$ is a submanifold and ω_0, ω_1 are two closed 2-forms on M for which $(\omega_0)|_N = (\omega_1)|_N$. Then there is a neighbourhood U of N and a diffeomorphism $f : U \to U$ such that*

- *$f(n) = n$ for all $n \in N$,*
- *$f^*\omega_1 = \omega_0$.*

Moreover, if $\Phi : G \times M \to M$ is an action on M by a compact group preserving N and the symplectic forms ω_0 and ω_1, then f can be chosen to be equivariant with respect to G. This means that $f \circ \Phi_g = \Phi_g \circ f$ for all $g \in G$.

Before giving the proof of the theorem, we state an important example and a corollary.

© The Author(s), under exclusive licence to Springer Nature Switzerland AG 2019
S. Dwivedi et al., *Hamiltonian Group Actions and Equivariant Cohomology*,
SpringerBriefs in Mathematics,
https://doi.org/10.1007/978-3-030-27227-2_3

Example 3.3 Let (M, ω_1) be a symplectic manifold. Fix coordinates identifying a neighbourhood of 0 in $T_m M$ with a neighbourhood of m in M. Since $T_m M$ is just a vector space, it has a canonical symplectic form ω_0.

Now consider a symplectic action by a compact group G on M. The fixed point set of G is a symplectic submanifold of M.

If G fixes the point m, then the action of G on $T_m M$ is linear. The equivariant Darboux theorem says that

- There is a coordinate system on a neighbourhood of m with respect to which ω_0 is the standard antisymmetric form on a symplectic vector space and the action of G is linear.
- $\phi^* \omega_1 = \omega_0$.

In other words, there exist Darboux coordinates around m with respect to which the action of G is linear.

Generalizing this example, we have the following.

Corollary 3.4 *Suppose $F \subset M$ is a submanifold that is fixed by an acting group G. By the tubular neighbourhood theorem, there exists an open subset U of the normal bundle $\nu(F)$ of F in M, containing the zero section of F, that embeds in M, and for which G acts linearly on the fibres of $\nu(F)$. That is, Darboux coordinates may be chosen near F for which the action of G is linear on the fibres of the normal bundle to F.*

We now give the proof of the equivariant Darboux–Weinstein theorem.

Proof This proof uses Moser's method. Consider

$$\omega_t = (1 - t)\omega_0 + t\omega_1.$$

For all t, ω_t is closed since both ω_0 and ω_1 are. Since $d(\omega_0 - \omega_1) = 0$, we can find a 1-form β such that $d\beta = \omega_0 - \omega_1$, in a neighbourhood of N. If N is a point, then we can choose a contractible neighbourhood of N and the result is obvious. Otherwise, we choose an equivariant family of maps $\phi_t : U \to U$ such that

(a) ϕ_t fixes N

(b) $\phi_0 : U \to N$, $\phi_1 = \text{id}$.

If X is a tubular neighbourhood of N identified with the normal bundle $\nu(N)$, then for $x = (y, v)$, $y \in N$, $v \in \nu(N)$, we have $\phi_t(y, v) = tv$.

Then for any form σ on M,

$$\phi_1^* \sigma - \phi_0^* \sigma = \int_0^1 \frac{d}{dt}\left(\phi_t^* \sigma\right) dt$$

$$= \int_0^1 \phi_t^*(L_{\xi_t}\sigma) dt$$

$$= \int_0^1 \phi_t^*\left(d i_{\xi_t}\sigma + i_{\xi_t} d\sigma\right) dt$$

$$= I d\sigma + dI\sigma$$

where we have defined a chain homotopy

$$I : \Omega^*(M) \to \Omega^{*-1}(M)$$

with

$$I\sigma = \int_0^1 \phi_t^* \left(i_{\xi_t}\sigma \right) dt.$$

Choosing $\sigma = \omega_0 - \omega_1$, we see that $d\sigma = 0$ in some neighbourhood $Y \subset N$ and

$$\beta = I\sigma, \qquad \sigma = d\beta.$$

It follows that $\beta|_Y = 0$.

Since $\omega_t|_Y$ is non-degenerate for all $t \in [0, 1]$, it follows that this is also true on some suitably small neighbourhood of N. Then we can find a time-dependent vector field η_t such that

$$i_{\eta_t}\omega_t = \beta.$$

Note that β may be chosen to be invariant under G and thus so may the time-dependent vector field η_t.

Integrating the vector fields η_t gives a family of local diffeomorphisms f_t with $f_0 = \mathrm{id}$ and

$$\frac{d}{dt} f_t(m) = \eta_t(f_t(m)).$$

Since the vector field η_t commutes with the action of G, the maps f_t are G-equivariant. Also $(\eta_t)|_Y = 0$ and so $(f_t)|_Y = \mathrm{id}$. We then have

$$
\begin{aligned}
(f_1)^* \omega_1 - \omega_0 &= \int_0^1 \frac{d}{dt}(f_t^* \omega_t) dt \\
&= \int f_t^* d(i_{\eta_t}\omega_t) dt + \int f_t^*(\omega_0 - \omega_1) dt \\
&= \int f_t^* d(-\beta) dt + \int f_t^*(\omega_0 - \omega_1) dt \\
&= \int f_t^*(\omega_1 - \omega_0) dt + \int f_t^*(\omega_0 - \omega_1) dt \\
&= 0
\end{aligned}
$$

Thus f_1 is the desired equivariant diffeomorphism. $\square$

References

1. M. Audin, *Torus Actions on Symplectic Manifolds*, vol. 93, Progress in Mathematics (Birkhäuser, Birkhäuser, 2004)
2. V. Guillemin, S. Sternberg, *Symplectic Techniques in Physics* (Cambridge University Press, Cambridge, 1986)

Chapter 4
Elementary Properties of Moment Maps

If a Lie group acts on a symplectic manifold preserving the symplectic form, it is possible that each fundamental vector field is the Hamiltonian vector field of a Hamiltonian function called the moment map which was defined in Definition 2.9.

In this chapter, we describe elementary properties of moment maps. The layout of the chapter is as follows. The first section (Sect. 4.1) outlines basic properties of moment maps. The second section (Sect. 4.2) gives examples of moment maps. The final section (Sect. 4.3) gives the normal form for a moment map.

4.1 Introduction

The following Proposition is immediate from the definition of a moment map.

Proposition 4.1 *1. Let $H \xrightarrow{p} G$ be a homomorphism of Lie groups, and let*

$$p^* : \mathbf{g}^* \to \mathbf{h}^*$$

be the obvious projection. Suppose G acts on M in a Hamiltonian fashion. Then the induced action of H on M is also Hamiltonian, with moment map $\Phi_H = p^* \circ \Phi_G$.
2. *In particular, this applies if H is a Lie subgroup of G and p is the inclusion map.*
3. *If M_1 and M_2 are two symplectic manifolds equipped with Hamiltonian actions of G, with moment maps $\Phi_1 : M_1 \to \mathbf{g}^*$ and $\Phi_2 : M_2 \to \mathbf{g}^*$, then the induced action on $M_1 \times M_2$ is also Hamiltonian with moment map $\Phi_1 + \Phi_2$.*
4. *If G and H both act on M in a Hamiltonian fashion with moment maps Φ_G and Φ_H and these actions commute, then $G \times H$ acts on M and the moment map is $\Phi_G \oplus \Phi_H : M \to \mathbf{g}^* \oplus \mathbf{h}^*$.*

Remark 4.2 Two flows (coming from group actions) commute if and only if the vector fields commute. This happens if and only if the Hamiltonian functions Poisson

© The Author(s), under exclusive licence to Springer Nature Switzerland AG 2019
S. Dwivedi et al., *Hamiltonian Group Actions and Equivariant Cohomology*,
SpringerBriefs in Mathematics,
https://doi.org/10.1007/978-3-030-27227-2_4

commute. So two flows each generating $U(1)$ actions fit together to form a $U(1) \times U(1)$ action if and only if their moment maps Poisson commute.

Proposition 4.3 ([1]) *Let $\Phi : M \to \mathbf{g}^*$ be a moment map so $d\Phi_m : T_m M \to \mathbf{g}^*$, for all $m \in M$. Then $\mathrm{Im}(d\Phi_m)^\perp = \mathrm{Lie}(\mathrm{Stab}(m))$ where $\perp$ denotes the annihilator under the pairing $\mathbf{g}^* \otimes \mathbf{g} \to \mathbb{R}$.*

Proof We have

$$i_{Y^\#}\omega = d\Phi_Y.$$

Thus Y annihilates all $\xi \in \mathrm{Im}(d\Phi_m)$ if and only if $Y \in \mathrm{Lie}(\mathrm{Stab}(m))$. $\square$

Corollary 4.4 *Zero is a regular value of Φ if and only if $\mathrm{Stab}(m)$ is finite for all $m \in \Phi^{-1}(0)$. In this situation, $\Phi^{-1}(0)$ is a manifold and the action of G on it is locally free.*

Example 4.5 Let T be a torus acting on M, and let $F \subset M^T$ be a component of the fixed point set. Then for any $f \in F$ we have $d\Phi_f = 0$ so $\Phi(F)$ is a point.

Proposition 4.6

- *If $H \subset G$ are two groups acting in a Hamiltonian fashion on a symplectic manifold M, then $\Phi_H = \pi \circ \Phi_G$ where $\pi : \mathbf{g}^* \to \mathbf{h}^*$ is the projection map. In other words if $X \in \mathbf{h}$ then $\Phi_H(m)(X) = \Phi_G(m)(X)$ for any $m \in M$. One example that frequently arises is the case when $H = T$ is a maximal torus of a compact Lie group G.*
- *More generally, if $f : H \to G$ is a Lie group homomorphism, and the two groups G and H act in a Hamiltonian fashion on a symplectic manifold M, in such a way that the action is compatible with the homomorphism f, then $\mu_H = f^* \circ \mu_G$ where $f^* : \mathbf{g}^* \to \mathbf{h}^*$ is induced from the homomorphism f. (The previous case is the special case where f is the inclusion map.)*
- *If two symplectic manifolds M_1 and M_2 are acted on in a Hamiltonian fashion by a group G with moment maps Φ_1 and Φ_2, then the moment map for the diagonal action of G on $M_1 \times M_2$ with the product symplectic structure is $\Phi_1 + \Phi_2$.*

4.2   Examples of Moment Maps

Example 4.7 Consider the space $M = \mathbb{R}^2 = \{(p, x)\}$. A symplectic form is defined on M by

$$\omega = dx \wedge dp.$$

This space is phase space for one degree of freedom. The variables x and p denote position and momentum, respectively.

The additive group $\mathbb{R}$ acts on $\mathbb{R}^2$ by $t.(p, x) \mapsto (p, x + t)$. where $t \in \mathbb{R}$. This action preserves ω. The vector field generated by it is $X := \frac{\partial}{\partial x}$. So

$$i_X(dx \wedge dp) = dp$$

and the moment map for the action of $\mathbb{R}$ on phase space is

$$\Phi(x, p) = p.$$

A general principle in physics—*Noether's theorem*—specifies that for any symmetry group of a physical system, there is a conserved quantity. The conserved quantity corresponding to symmetry under translation is linear momentum. The conserved quantity corresponding to symmetry under rotation is angular momentum.

Example 4.8 Consider the space

$$M = \mathbb{R}^6 = \{(p_1, p_2, p_3, x_1, x_2, x_3)\}.$$

Define $\bar{p} = (p_1, p_2, p_3)$ and $\bar{x} = (x_1, x_2, x_3)$. We will examine the Hamiltonian flow of the function

$$\Phi(\bar{x}, \bar{p}) = \bar{x} \times \bar{p}$$

where $\bar{x} \times \bar{p}$ is the cross product

$$\bar{x} \times \bar{p} = \sum_{i,j,k} \epsilon_{ijk} x_j p_k \hat{e}_i$$

where $\{\hat{e}_j | j = 1, 2, 3\}$ is the standard basis of $\mathbb{R}^3$. Here $\epsilon_{123} = 1$, $\epsilon_{213} = -1$ (and cyclic permutations) and $\epsilon_{ijk} = 0$ if any two of i, j, k are equal. So

$$d\Phi_{\hat{e}_i} = \sum_{i,j,k} \epsilon_{ijk}(dx_j p_k + x_j dp_k)$$

Then

$$i_{\frac{\partial}{\partial x_i}} \omega = dp_i,$$

$$i_{\frac{\partial}{\partial p_i}} \omega = -dx_i.$$

So

$$X_{\hat{e}_i} = \sum_{j,k} \epsilon_{ijk} \left(p_j \frac{\partial}{\partial p_k} + x_j \frac{\partial}{\partial x_k} \right)$$

This is the vector field generated by the action of $\hat{e}_i \in so(3)$ on $\mathbb{R}^6$.

Example 4.9 Let $U(n)$ act on $\mathbb{C}^n$ by component-wise multiplication of complex numbers. Since $\mathfrak{u}(n) \cong \mathbb{R}^n$, $\phi : \mathbb{C}^n \to \mathbb{R}^n$. For $X \in \text{Lie}(U(n))$ and

$$z = (z_1, \ldots, z_n) \in \mathbb{C}^n,$$

the moment map for this action is

$$\Phi(z)(X) = \frac{1}{2\pi} z^T X z$$

$$= \frac{1}{2\pi} \sum_{i,j=1}^{n} z_i X_{ij} z_j.$$

Example 4.10 Let $U(n)$ act on $\mathbb{C}P^{n-1}$ via its action on $\mathbb{C}^n$. The moment map is then

$$\Phi(z_1, \ldots, z_n)(X) = \frac{1}{2\pi |z|^2} \left(z^T X z \right).$$

This will follow from the previous example once we have introduced symplectic quotients in Chap. 6. This tells you the moment map for any linear action of a compact group G on $\mathbb{C}^n$ or $\mathbb{C}P^{n-1}$.

Example 4.11 Let V be a complex vector space with an action of a torus T. Recall the weight lattice $\Lambda^W = \text{Hom}(T, U(1))$. (For more information see Chap. 11 and Appendix.) Decompose

$$V = \mathbb{C}_{\lambda_1} \oplus \ldots \oplus \mathbb{C}_{\lambda_n}$$

where T acts on $\mathbb{C}_{\lambda_j}$ via the weight $\exp(2\pi i \lambda_j) : T \to U(1)$, where $\lambda_j : \mathbf{t} \to \mathbb{R}$ sends the kernel of the exponential map to $\mathbb{Z}$.

Example 4.12 (a) The moment map Φ_λ for action of T on $\mathbb{C}_\lambda$:

$$p_j : u(1)^* \to t^*.$$

The map $\Phi_\lambda(z)$ is the composition of $\Phi_{U(1)} : \mathbb{C} \to \mathbb{R}$, defined by

$$\Phi_{U(1)}(z) = \frac{|z|^2}{2}$$

with $p_j : \mathbb{R} \to t^*$ given by

$$p_j(s) = s\lambda_j.$$

So

$$\Phi_j(z) = -\frac{1}{2} |z|^2 \lambda_j.$$

(b) The moment map for the action of T on $\oplus_j \mathbb{C}_{\lambda_j}$ is

$$\Phi = \sum_j \Phi_{\lambda_j}$$

$$\Phi(z_1, \ldots, z_n) = -\frac{1}{2} \sum_j |z_j|^2 \lambda_j$$

(c) The moment map for action of $U(1)$ on $\mathbb{C}^n$ is

$$\Phi(z_1, \ldots, z_n) = -\frac{1}{2} \sum_j |z_j|^2.$$

Example 4.13 $SU(2)$ action on $(\mathbb{C}P^1)^N$
For $SU(2)$ acting on $\mathbb{C}P^1 = S^2 \subset \mathbb{R}^3$, the moment map is the inclusion map

$$\Phi = i : S^2 \to \mathbb{R}^3.$$

For $SU(2)$ action on $(\mathbb{C}P^1)^N$, the moment map is

$$\Phi(z_1, \ldots, z_N) = z_1 + \ldots + z_N.$$

4.3 The Normal Form Theorem

We denote by M_{red} the reduced space (or the symplectic quotient) $\phi^{-1}(0)/G$ and by ω_{red}, the symplectic form on M_{red}. (The symplectic quotient and its symplectic form will be defined in Chap. 6.) There is a neighbourhood of $\Phi^{-1}(0)$ on which the symplectic form is given in a standard way related to the symplectic form ω_{red} on M_{red} (see, for instance, Sects. 39–41 of [2]). Before stating the theorem, recall that a connection on a principal G-bundle is a Lie algebra-valued 1-form, i.e., if θ is a connection on a principal G-bundle P, then $\theta \in \Omega^1(P) \otimes \mathbf{g}$. Also, denote by $i : M_{\mathrm{red}} \to M$ the inclusion map. We now state the follows.

Proposition 4.14 ([2, Proposition 40.1] (Normal form theorem)) *Assume 0 is a regular value of Φ (so that $\Phi^{-1}(0)$ is a smooth manifold and G acts on $\Phi^{-1}(0)$ with finite stabilizer). Then there is a neighbourhood $\mathcal{O} \cong \Phi^{-1}(0) \times \{z \in \mathbf{g}^*, |z| \le \epsilon\} \subset \Phi^{-1}(0) \times \mathbf{g}^*$ of $\Phi^{-1}(0)$ on which the symplectic form is given as follows. Let $P \stackrel{def}{=} \Phi^{-1}(0) \stackrel{q}{\to} M_{\mathrm{red}}$ be the orbifold principal G-bundle given by the projection map $q : \Phi^{-1}(0) \to \Phi^{-1}(0)/G$, and let $\theta \in \Omega^1(P) \otimes \mathbf{g}$ be a connection for it. Let ω_0 denote the induced symplectic form on M_{red}, in other words $q^*\omega_0 = i_0^*\omega$. Then if we define a 1-form τ on $\mathcal{O} \subset P \times \mathbf{g}^*$ by $\tau_{p,z} = z(\theta)$ (for $p \in P$ and $z \in \mathbf{g}^*$), the symplectic form on $\mathcal{O}$ is given by*

$$\omega = q^*\omega_0 + d\tau.$$

Further, the moment map on $\mathcal{O}$ is given by $\Phi(p, z) = z$.

Corollary 4.15 *Suppose that t is a regular value for the moment map for the Hamiltonian action of a torus T on a symplectic manifold M. Then in a neighbourhood of $\mathbf{g}$, all symplectic quotients M_t are diffeomorphic to M_{t_0} by a diffeomorphism under which*

$$\omega_t = \omega_{t_0} + (t - t_0, d\theta)$$

where $\theta \in \Omega^1\left(\Phi^{-1}(t_0)\right) \otimes \mathbf{t}$ is a connection for the action of T on $\Phi^{-1}(t_0)$.

Corollary 4.16 *Suppose G acts in a Hamiltonian fashion on a symplectic manifold M, and suppose 0 is a regular value for the moment map Φ. Then the reduced space $M_\lambda = \Phi^{-1}(\mathcal{O}_\lambda)/G$ at the orbit $\mathcal{O}_\lambda$ fibres over $M_0 = \Phi^{-1}(0)/G$ with fibre the orbit $\mathcal{O}_\lambda$. Furthermore, if $\pi : M_\lambda \to M_0$ is the projection map, then the symplectic form ω_λ on $\Phi^{-1}(\mathcal{O}_\lambda)/G$ is given as $\omega_\lambda = \pi^*\omega_0 + \Omega_\lambda$ where ω is the symplectic form on M_0 and Ω_λ restricts to the standard Kirillov–Kostant–Souriau symplectic form on the fibre. (See Chap. 5.)*

References

1. V. Guillemin, S. Sternberg, Convexity properties of the moment mapping I and II. Invent. Math. **67**, 491–513 (1982); **77**, 533–546 (1984)
2. V. Guillemin, S. Sternberg, *Symplectic Techniques in Physics* (Cambridge University Press, Cambridge, 1986)

Chapter 5
The Symplectic Structure on Coadjoint Orbits

In this chapter, we explain why the orbit of the adjoint action on the Lie algebra of a Lie group is symplectic, and define its symplectic form (the Kirillov–Kostant–Souriau form). An example of an orbit of the adjoint action is the two-sphere, which is an orbit of the action of the rotation group $SO(3)$ on its Lie algebra $\mathbb{R}^3$. Background information on Lie groups may be found in Appendix.

A Lie group G acts smoothly on its dual Lie algebra $\mathbf{g}^*$ through the coadjoint action. Given an element $X \in \mathbf{g}$, it generates a vector field $X^\sharp$ on $\mathbf{g}^*$ whose value at $\lambda \in \mathbf{g}^*$ is $X_\lambda^\sharp = [\lambda, X]$, where by definition $(X^\sharp)_\lambda(Y) = \lambda([X, Y])$ for $Y \in \mathbf{g}$. It follows that the tangent space to O_λ at λ is $\{[\lambda, X] : X \in \mathbf{g}\}$. We can give the coadjoint orbit a symplectic structure by introducing the Kirillov–Kostant–Souriau 2-form, defined on O_λ by

$$\omega_\lambda(X_\lambda^\sharp, Y_\lambda^\sharp) = -\lambda([X, Y]).$$

This 2-form is indeed symplectic.

Theorem 5.1 *Suppose $\lambda \in \mathbf{g}^*$ and O_λ is the coadjoint orbit through λ. Then O_λ carries a symplectic structure, ω_λ, called the Kirillov–Kostant–Souriau form. This was defined above. It satisfies*

1. *ω_λ is preserved by the action of G.*
2. *ω_λ is closed.*
3. *ω_λ restricts to a non-degenerate form on the coadjoint orbit O_λ.*
4. *The moment map for the action of G on the orbit is given by the inclusion map $O_\lambda \to \mathbf{g}^*$.*

Proof We prove the four items in order:

1. Recall that for $X \in \mathbf{g}$ and $g \in G$ we have that $\mathrm{Ad}_{g^{-1},*}(X^\sharp) = (\mathrm{Ad}_g(X))^\sharp$. It follows that

© The Author(s), under exclusive licence to Springer Nature Switzerland AG 2019
S. Dwivedi et al., *Hamiltonian Group Actions and Equivariant Cohomology*,
SpringerBriefs in Mathematics,
https://doi.org/10.1007/978-3-030-27227-2_5

$$(\mathrm{Ad}^*_{g^{-1}}\omega)_\lambda(X^\sharp, Y^\sharp) = \omega_{\mathrm{Ad}^*_{g^{-1}}\lambda}(\mathrm{Ad}_{g^{-1},*}(X^\sharp), \mathrm{Ad}_{g^{-1},*}(Y^\sharp))$$
$$= -\mathrm{Ad}^*_{g^{-1}}\lambda\left([\mathrm{Ad}_{g,*}(X), \mathrm{Ad}_{g,*}(Y)]\right)$$
$$= -\mathrm{Ad}^*_{g^{-1}}\lambda(\mathrm{Ad}_{g,*}[X, Y])$$
$$= -\lambda([X, Y])$$
$$= \omega_\lambda(X, Y).$$

2. We have that

$$d\omega_\lambda(X^\sharp, Y^\sharp, Z^\sharp) =$$

$$\frac{1}{3}\left\{X^\sharp(\omega_\lambda(Y^\sharp, Z^\sharp)) - Y^\sharp(\omega_\lambda(X^\sharp, Z^\sharp)) + Z^\sharp(\omega_\lambda(X^\sharp, Y^\sharp))\right\} +$$

$$\frac{1}{3}\left\{-\omega([X^\sharp, Y^\sharp], Z^\sharp) + \omega([X^\sharp, Z^\sharp], Y^\sharp) - \omega([Y^\sharp, Z^\sharp], X^\sharp)\right\}.$$

For the first bracket, by (1) it follows that $(L_{X^\sharp}\omega)(Y^\sharp, Z^\sharp) = 0$. Hence, by definition of the Lie derivative,

$$X^\sharp(\omega(Y^\sharp, Z^\sharp)) = \omega([X^\sharp, Y^\sharp], Z^\sharp) + \omega(Y^\sharp, [X^\sharp, Z^\sharp]).$$

Applying the same argument to $Y^\sharp(\omega(X^\sharp, Z^\sharp))$ and $Z^\sharp(\omega(X^\sharp, Y^\sharp))$, and adding shows that the first bracket vanishes. For the second bracket notice that

$$\omega_\lambda([X^\sharp, Y^\sharp], Z^\sharp) = \omega_\lambda([X, Y]^\sharp, Z^\sharp) = -\lambda([[X, Y], Z]).$$

Hence the second bracket becomes

$$\lambda([[X, Y], Z] + [[Z, X], Y] + [[Y, Z], X])$$

which vanishes by the Jacobi identity.
3. To show that ω_λ is non-degenerate, consider an arbitrary element $[X, \lambda] \in T_\lambda O_\lambda$ satisfying

$$\lambda([X, \cdot]) = 0.$$

Now, the elements of **g** whose infinitesimal generator vanishes at λ are precisely the elements in $\mathbf{g}_\lambda$, the Lie algebra of the stabilizer of λ. Hence, we see that the kernel of ω_λ is $\mathbf{g}_\lambda$, the Lie algebra of the stabilizer of λ. But since $T_\lambda O_\lambda$ can be identified with $\mathbf{g}/\mathbf{g}_\lambda$ it follows that ω_λ is non-degenerate on $T_\lambda O_\lambda$.
4. Let $\iota : O_\lambda \to \mathbf{g}^*$ be the inclusion mapping. Then for $Y \in \mathbf{g}$ we have

$$\iota_*([X, \lambda])(Y) = [X, \lambda](Y)$$
$$= -\lambda([X, Y])$$
$$= \omega_\lambda([X, \lambda], [Y, \lambda]).$$

Hence, by definition, the inclusion map is the moment map.   □

If we write the above 2-form in terms of $\lambda \in \mathbf{g}^*$, it is canonical; in other words, it does not depend on a choice of inner product on $\mathbf{g}$. However, it is often more convenient to choose an inner product $\langle \cdot, \cdot \rangle$ on $\mathbf{g}$, to identify $\mathbf{g}^*$ with $\mathbf{g}$. Then, we can work with the adjoint action rather than the coadjoint action. Once such an inner product has been chosen, we can write the 2-form as

$$\omega_\lambda(\hat{X}, \hat{Y}) = -\langle \lambda, [X, Y] \rangle.$$

This 2-form depends on the choice of inner product.

Remark 5.2 There is an inner product on G which is invariant under both left and right multiplication. This inner product gives rise to an inner product on $\mathbf{g}$ which is invariant under the adjoint action of G. For simple Lie algebras, such an inner product on $\mathbf{g}$ is unique up to multiplication by a constant. For semisimple Lie algebras, the inner product is unique up to multiplication by constants on all the simple factors. One example of such an inner product is called the *Killing form*. For more information, see [1].

Remark 5.3 Each O_λ is isomorphic to $G/Z(\lambda)$ where $Z(\lambda)$ is the centralizer of λ. The centralizer will always include some maximal torus; however, for most λ, the centralizer $Z(\lambda)$ is just one maximal torus. For instance, if $G = U(n)$ and λ is a diagonal matrix with no equal eigenvalues, then the subgroup of G commuting with λ is just T, the diagonal matrices $U(1)^n$. For some λ, $Z(\lambda)$ is larger, for instance, if there are some equal eigenvalues.

Reference

1. T. Bröcker, T. Tom Dieck, *Representations of Compact Lie Groups*, GTM (Springer, Berlin, 1985)

Chapter 6
Symplectic Reduction

6.1 Introduction

In the presence of a Hamiltonian action on a symplectic manifold, we may reduce the size of the symplectic structure by quotienting out by the symmetries. When and how we can do this is the primary subject of this chapter. Sufficient conditions are described by the theorem of Marsden–Weinstein and Meyer. The rest of the chapter is devoted to studying various equivalent definitions of symplectic quotients (Sect. 6.2), the shifting trick (Sect. 6.3), the structure of symplectic quotients by subgroups (Sect. 6.4) and finally constructing symplectic cuts (Sect. 6.5), which will be a useful tool in later chapters. Many of the results and proofs found in this chapter rely on Ana Cannas da Silva's book [1].

Let us start by recalling some standard definitions and results on smooth manifolds which are covered for instance in Lee's book on smooth manifolds [2].

Definition 6.1 A smooth map $f : M \to N$ is said to be *transverse* to a submanifold $S \subset N$ when for every $x \in f^{-1}(S)$ the tangent space $T_{f(x)}N$ is spanned by the subspaces $T_{f(x)}S$ and $df_x(T_x M)$.

Theorem 6.2 *If $f : M \to N$ is a smooth map which is transverse to an embedded submanifold $S \subset N$, then $f^{-1}(S)$ is an embedded submanifold of M with codimension in M equal to that of S in N. In particular, if $S = \{y\}$ for a regular value y of f then $f^{-1}(S)$ is an embedded submanifold of M with codimension equal to the dimension of N.*

Regular level sets provide a wealth of examples of embedded submanifolds and we might hope that by considering certain functions on a symplectic manifold (M, ω) the regular levels with inherit a natural symplectic structure. If (M, ω) comes with a Hamiltonian G-action, then the regular levels of the moment map are a useful subject of study. However, because the moment map is constant on the orbits it follows that

© The Author(s), under exclusive licence to Springer Nature Switzerland AG 2019
S. Dwivedi et al., *Hamiltonian Group Actions and Equivariant Cohomology*,
SpringerBriefs in Mathematics,
https://doi.org/10.1007/978-3-030-27227-2_6

the restriction of ω to a regular level will be degenerate along any direction which is also tangent to the orbit,

$$\iota_{X^\#} i^*\omega = -i^*(d\mu^X) \equiv 0 \qquad \forall\, X^\# \in T_x(G \cdot x) \bigcap T_x\big(\mu^{-1}(\xi)\big).$$

On each tangent space $T_x\mu^{-1}(\xi)$, we can remove the degeneracies of the linear form $i^*\omega_x$ by quotienting out the infinitesimal symmetries. Then we hope that $i^*\omega_x$ will push forward to a non-degenerate linear form on this quotient space. As the infinitesimal symmetries are the tangent vectors to the G-orbits, we can hope that the action of G will extend this fibrewise notion to a global one. To pursue this approach, we must first recall some definitions and results surrounding principal bundles [3].

Definition 6.3 An action $G \xrightarrow{\psi} \mathrm{Diff}(M)$ by a Lie group G on a smooth manifold M is said to be *free* if the stabilizer of every point is trivial and is said to be *locally free* if the stabilizer of every point is finite. The action is said to be *proper* when the map $G \times M \to M \times M$ by $(g, x) \mapsto (\psi_g(x), x)$ is a proper map. Note that if G is assumed to be compact then the action is automatically proper.

Definition 6.4 (*Fibre Bundles*) A *fibre bundle* is a map $p : M \to B$ of smooth manifolds which is locally a projection, that is, there is an open covering of B by sets U_i and diffeomorphisms $\phi_i : p^{-1}(U_i) \to U_i \times E$ so that $p : p^{-1}(U_i) \to U_i$ is the composition of ϕ_i with the projection onto the first factor. The spaces M, B and E are called the *total space*, *base space* and *typical fibre* of the fibre bundle ($p : M \to B$), respectively, and we refer to the preimages of points $p^{-1}(y)$ as the *fibres*. On any fibre bundle, there is a natural short exact sequence of vector bundles obtained by differentiating the map p, and the kernel of Tp is the canonically defined *vertical bundle* which we denote by $VM = \ker Tp$,

$$0 \longrightarrow VM \longrightarrow TM \xrightarrow{Tp} p^*TB \longrightarrow 0 .$$

Definition 6.5 (*Principal Bundles*) A *principal G-bundle with structure group G* is a fibre bundle $p : M \to B$ with a right action of G on the total space which is also free and transitive on the fibres, that is

$$\forall x \in M \; g \in G \qquad g \cdot x \in p^{-1}(p(x))$$
$$\forall x \in M \quad g \cdot x = x \;\Rightarrow\; g = 1$$
$$\forall x, y \in M \quad p(x) = p(y) \;\Rightarrow\; \exists g \in G \; g \cdot x = y.$$

The fibres of the bundle are therefore diffeomorphic to G and so the base B may be identified with the orbit space M/G.

Proposition 6.6 *If a Lie group G acts freely and properly on a manifold M, then the orbit space M/G is a smooth manifold and the orbit map $p : M \to M/G$ is a principal G-bundle.*

Definition 6.7 (*Horizontal and basic forms*) A *horizontal form* on a fibre bundle is a differential form on the total space such that the contraction with any vertical vector field vanishes. A *basic form* on a principal bundle is an invariant horizontal form.

$$\Omega^k_{hor}(p : M \to B) = \left\{\alpha \in \Omega^k(M) \middle| \iota_X\alpha \equiv 0 \;\; \forall X \in VM\right\}$$

$$\Omega^k_{bas}(p : M \to B, G) = \left\{\alpha \in \Omega^k_{hor}(p : M \to B) \middle| \psi_g^*\alpha = \alpha \;\; \forall g \in G\right\}.$$

The reason for considering basic forms on principal bundles is that they are precisely those forms on the total space that can be pushed forward to the base.

Proposition 6.8 *Every basic k-form* $\alpha \in \Omega^k_{bas}(P)$ *on a principal G-bundle* $p : M \to B$ *determines a unique k-form* $p_*\alpha \in \Omega^k(B)$ *which is a pushforward of* α *in the sense that* $p^*p_*\alpha = \alpha$. *Moreover, if* α *is closed then so is the pushforward* $p_*\alpha$.

We now have all the tools to make precise our vague description of quotienting regular levels to obtain a symplectic structure. Before doing so in generality we exhibit the basic idea through the example of the usual $U(1)$ action on $\mathbb{C}^{n+1}$.

Example 6.9 (*Fubini–Study structure on complex projective space*) Consider the complex vector space $\mathbb{C}^{n+1}$ with coordinates $(z_0, \ldots, z_n)$ along with the standard symplectic structure and the usual Hamiltonian action of $U(1)$ with associated vector field ∂_θ and moment map μ,

$$\omega = \frac{1}{2i}\sum_{j=0}^{n} dz_j \wedge d\bar{z}_j, \qquad \partial_\theta = \sum_{j=0}^{n} i\left(z_j\partial_{z_j} - \bar{z}_j\partial_{\bar{z}_j}\right) \qquad \mu(z) = \frac{1}{2}\|z\|^2.$$

Any nonzero level of μ is a codimension 1 sphere and we denote by

$$i : S^{2n+1} \hookrightarrow \mathbb{C}^{n+1}$$

the inclusion of the unit sphere $\mu^{-1}(1/2)$. We can represent the restriction of ω to S^{2n+1} explicitly as a 2-form which vanishes on the normal bundle NS^{2n+1} and agrees with ω on vectors tangent to S^{2n+1}; to this end consider the 2-form $\omega + d\mu \wedge d\theta$. Showing that this 2-form vanishes on the normal bundle amounts to checking that it vanishes on the global Euler vector field ∂_e which generates NS^{2n+1}. See (6.1) for the definition of this vector field.

$$\iota_{\partial_e}d\mu \equiv 1 \quad \rightsquigarrow \quad \partial_e = \sum_{j=0}^{n}(z_j\partial_{z_j} + \bar{z}_j\partial_{\bar{z}_j}) \qquad \iota_{\partial_e}\omega = \sum_{j=0}^{n} \frac{z_j d\bar{z}_j - \bar{z}_j dz_j}{2i} \qquad (6.1)$$

$$= -\|z\|^2 d\theta = -d\theta.$$

Since $d\theta$ vanishes on ∂_e, it follows that $\iota_{\partial_e}(d\mu \wedge d\theta) = -d\theta = -\iota_{\partial_e}\omega$ and so $\iota_{\partial_e}(\omega + d\mu \wedge d\theta) = 0$ as desired. On the other hand $\omega + d\mu \wedge d\theta$ agrees with ω

whenever contracted with a vector field tangent to S^{2n+1}. Indeed $\iota_{\partial_\theta}\omega = \iota_{X^\#}\omega = -d\mu^X = -Xd\mu = -d\theta(\partial_\theta)d\mu = d\mu \wedge d\theta$. Locally complete $\partial_\mathbf{e}, \partial_\theta$ to a basis so that the $d\mu \wedge d\theta$ vanishes for tangent vectors orthogonal to the orbits.

The 2-form $\omega + d\mu \wedge d\theta$ is certainly closed but fails to be symplectic when restricted to S^{2n+1} since ∂_θ is tangent to S^{2n+1} but as we have already seen $\iota_{\partial_\theta}(\omega + d\mu \wedge d\theta) = 2d\mu$ and $d\mu$ vanishes on $TS^{2n+1} = \ker d\mu$ by definition. This makes $\omega + d\mu \wedge d\theta$ a horizontal form for the principal $U(1)$-bundle

$$p : S^{2n+1} \to \mathbb{C}P^n,$$

and since both ω and $d\mu \wedge d\theta$ are $U(1)$-invariant $\omega + d\mu \wedge d\theta$ is in fact basic. We therefore have a well-defined pushforward of $\omega + d\mu \wedge d\theta$ to the base $\mathbb{C}P^n$ which we denote by ω^{FS}.

The closed 2-form ω^{FS} is in fact symplectic since by definition we have $\iota_{p_*X}\omega^{FS} = \iota_X(\omega + d\mu \wedge d\theta)$ and the latter vanishes if and only if X is a multiple of ∂_θ in which case p_*X is zero. The resulting symplectic structure $(\mathbb{C}P^n, \omega^{FS})$ is called the Fubini–Study structure on complex projective space and realizing it as above allows us to express it in coordinates as below:

Example 6.10

$$p^*\omega^{FS} = \frac{1}{\frac{1}{2}\|z\|^2}(\omega + d\mu \wedge d\theta) = \frac{1}{i\|z\|^2}\left(\sum_{j=0}^{n} dz_j \wedge d\bar{z}_j\right) \qquad (6.2)$$

$$-\frac{1}{4i\|z\|^4}\left(\left(\sum_{j=0}^{n} z_j d\bar{z}_j + \bar{z}_j dz_j\right) \wedge \left(\sum_{k=0}^{n} \bar{z}_k dz_k - z_k d\bar{z}_k\right)\right)$$

$$= \frac{1}{i\|z\|^4}\sum_{k=0}^{n}\sum_{j\neq k} \left(z_j\bar{z}_j dz_k \wedge dz_k - z_j\bar{z}_k dz_k \wedge d\bar{z}_j\right).$$

6.2 Symplectic Quotients

For this section, let us fix a connected symplectic manifold (M, ω) equipped with a Hamiltonian action of a compact and connected Lie group G having moment map μ. As in our example in the previous section we will want to consider the restriction of ω to a regular level of μ and then remove the degeneracies by pushing forward to the space of G-orbits. The latter step is essentially a fibrewise notion and it will be useful to isolate the linear version of this process after recalling some basic definitions from linear symplectic spaces.

Definition 6.11 Let (V, Ω) be a symplectic vector space and $F \subset V$ a subspace. The *symplectic orthogonal* or *symplectic annihilator* of F is the subspace

$$F^\Omega = \{v \in V \; : \; \Omega(v, u) = 0 \; \forall u \in F\}.$$

The subspace F is said to be *isotropic* if $F \subset F^\Omega$ and *coisotropic* if $F^\Omega \subset F$. Lastly, we say that F is a *Lagrangian* subspace if it is both isotropic and coisotropic; in other words, if it is equal to its symplectic orthogonal $F = F^\Omega$.

Lemma 6.12 (Linear reduction) *Suppose F is a subspace of a symplectic vector space (V, Ω) with inclusion $I : F \to V$. Denote by P the projection of F onto the quotient $\overline{F} = F/(F \cap F^\Omega)$. Then there exists a unique linear symplectic form Ω^F on $\overline{F}$ satisfying $P^*\Omega^F = I^*\Omega$.*

Proof The pullback relation forces us to define Ω^F as follows and, in particular, uniqueness is immediate,

$$\Omega^F([u], [v]) = \Omega(u, v) \qquad u, v \in F.$$

We can check that Ω^F is well-defined one argument at a time in which case this follows from the universal property of the kernel since Ω^F is defined on equivalence classes modulo $\ker I^*\Omega = F^\Omega \cap F$. It is clear that Ω^F is bilinear. The fact that Ω^F is symplectic is again a result of the universal property of the kernel but we can also see this explicitly; for any $[u] \in \overline{F}$ we have

$$\Omega^F([u], \overline{F}) = 0 \quad \Rightarrow \quad \Omega(u, F) = 0 \quad \Rightarrow \quad u \in F^\Omega \quad \Rightarrow \quad [u] = 0. \qquad \square$$

Unlike in our example, it is possible that the action of G may not preserve the levels of μ; so to obtain our principal bundle, we therefore consider the action of subgroups which do preserve $\mu^{-1}(\xi)$. Equivariance of μ implies that those elements of G which do preserve $\mu^{-1}(\xi)$ are precisely those that fix ξ under the coadjoint action. So the largest possible subgroup which acts on the level $\mu^{-1}(\xi)$ is the coadjoint stabilizer G_ξ of ξ. This means that we can at best remove the degenerate directions which are tangent to the orbit of G_ξ. Thankfully the following lemma ensures that this will be enough to recover a non-degenerate form from $i^*\omega$.

Lemma 6.13 *For any point $x \in \mu^{-1}(\xi)$, the symplectic orthogonal of the tangent space to $\mu^{-1}(\xi)$ at any $x \in \mu^{-1}(\xi)$ is the tangent space to the entire G-orbit through x. Moreover, the intersection of these subspaces in $T_x\mu^{-1}(\xi)$ is precisely the tangent space to the G_ξ orbit through x:*

$$\left(T_x\mu^{-1}(\xi)\right)^{\omega_x} = T_x(G \cdot x) \qquad T_x\left(\mu^{-1}(\xi)\right) \bigcap T_x(G \cdot x) = T_x\left(G_\xi \cdot x\right).$$

It follows from Lemma 6.12 that there is a family of uniquely determined (linear) symplectic forms ω_x^ξ on the quotients $T_x\mu^{-1}(\xi)/T_x(G_\xi \cdot x)$.

Proof The statement of the lemma amounts to the following equalities:

$$\begin{aligned}
\ker(i^*\omega_x) &= T_x\big(\mu^{-1}(\xi)\big) \cap \big(T_x\mu^{-1}(\xi)\big)^{\omega_x} \\
&= T_x\big(\mu^{-1}(\xi)\big) \cap T_x(G \cdot x) \\
&= T_x(G_\xi \cdot x).
\end{aligned}$$

The first is straight from the definition and the second can be verified by showing that $\ker(T_x\mu) = (T_x G \cdot x)^{\omega_x}$. Indeed, since μ is a moment map, we have for any $Y_x \in T_x M$

$$\langle (T_x\mu) Y_x, X \rangle = \iota_Y d\mu_x^X = \omega_x(X_x^\#, Y_x) \qquad \forall X \in \mathfrak{g}.$$

So $(T_x\mu)Y_x$ vanishes if and only if Y_x is ω_x-orthogonal to the span of the $X_x^\#$, which is to say that $T_x(\mu^{-1}(\xi)) = (T_x G \cdot x)^{\omega_x}$ and taking the symplectic orthogonal of both sides yields the equality. The last equality can be checked by writing out the intersection in terms of the infinitesimal action and using equivariance of $T_x\mu$,

$$T_x\mu^{-1}(\xi) \cap T_x(G \cdot x) = \left\{ X_x^\# \mid 0 = T_x\mu(X^\#) = \mathrm{ad}_X^* \xi \right\} \tag{6.3}$$

$$= \left\{ X_x^\# \mid \mathrm{Ad}_{\exp X}^* \xi = \xi \right\} \tag{6.4}$$

$$= \left\{ X_x^\# \mid X \in \mathfrak{g}_\xi \right\}. \tag{6.5}$$

$\square$

Now we are ready to present the main result of this chapter.

Theorem 6.14 (Symplectic reduction at a regular level) *Let (M, ω) be a connected symplectic manifold endowed with a Hamiltonian action of a connected Lie group G having moment map μ. For a point $\xi \in \mathfrak{g}^*$ let $i : \mu^{-1}(\xi) \to M$ denote the inclusion, and M^ξ the orbit space $\mu^{-1}(\xi)/G_\xi$ with corresponding projection $p : \mu^{-1}(\xi) \to M^\xi$. If ξ is a regular value of μ and G_ξ acts freely and properly on $\mu^{-1}(\xi)$, then there is a unique symplectic structure ω^ξ on M^ξ satisfying $p^*\omega^\xi = i^*\omega$.*

Proof Since ξ is assumed to be a regular value, the level $\mu^{-1}(\xi)$ is a smooth submanifold of M with codimension $k = \dim G$ and requiring the action of G_ξ on $\mu^{-1}(\xi)$ to be free and proper ensures the orbit mapping is a principal G_ξ-bundle. Then for any $x \in \mu^{-1}(\xi)$, we have short exact sequences induced by the inclusion and projection:

$$0 \longrightarrow T_x(G_\xi \cdot x) \longrightarrow T_x\mu^{-1}(\xi) \xrightarrow{T_x p} T_{p(x)}M^\xi \longrightarrow 0$$

$$0 \longrightarrow T_x\mu^{-1}(\xi) \xrightarrow{T_x i} T_x M \xrightarrow{T_x\mu} T_{\mu(x)}\mathfrak{g}^* \longrightarrow 0.$$

Now the pullback $i^*\omega$ is a smooth closed 2-form on $\mu^{-1}(\xi)$ and by our lemma on linear reduction, $(i^*\omega)_x$ vanishes precisely on $T_x(G_\xi \cdot x)$. So $i^*\omega$ is a closed basic 2-form on $\mu^{-1}(\xi)$ since it vanishes on the vertical bundle and G_ξ-equivariance follows from G-equivariance of ω and i. Therefore, there exists a unique pushforward to a closed 2-form ω^ξ on the base,

$$\omega^\xi \in \Omega^2(M^\xi) \qquad p^*\omega^\xi = i^*\omega.$$

This proves the uniqueness claim since any symplectic form on M^ξ satisfying the pullback relation is necessarily the pushforward of $i^*\omega$. For existence, it remains only to verify that ω^ξ is in fact symplectic which amounts to verifying non-degeneracy in the fibres. This too follows from the linear reduction lemma since $(\omega^\xi)_{p(x)}$ is the pushforward of the linear form $(i^*\omega)_x$ to the quotient of $T_x\left(\mu^{-1}(\xi)\right)$ by its kernel. □

6.3 Reduction at Coadjoint Orbits and the Shifting Trick

Equivariance of the moment map led us to consider the action of the coadjoint stabilizer on a regular level instead of the action of the entire group G. On the other hand, it follows from the equivariance of μ that we can consider an action of the entire group G if we are willing to enlarge the level ξ to the preimage of its entire coadjoint orbit $\mathcal{O}_\xi$. The following lemma shows us that the sufficient conditions to form the symplectic quotient (M^ξ, ω^ξ) as in the previous section are enough to ensure the action of G on $\mu^{-1}(\mathcal{O}_\xi)$ defines a principal G-bundle $p_{\mathcal{O}_\xi} : \mu^{-1}(\mathcal{O}_\xi) \to M^{\mathcal{O}_\xi}$.

Lemma 6.15 *If the coadjoint orbit $\mathcal{O}_\xi$ contains a single regular value of μ, then every point in $\mathcal{O}_\xi$ must be a regular value of μ. In this case, the moment map is transverse to the coadjoint orbit and hence $\mu^{-1}(\mathcal{O}_\xi)$ is a smooth submanifold of M of codimension equal to the codimension of G_ξ in G.*

Proof Assume a point $\xi \in \mathcal{O}_\xi$ is a regular value of μ so that $T_x\mu$ is surjective for all $x \in \mu^{-1}(\xi)$. By definition $T_x\mu_X = \omega_x(X^\#, \cdot)$ for any X in the Lie algebra and $x \in M$. This means that since we are assuming $T_x\mu$ is surjective, $X^\# \neq 0$ for all X. This implies that for $x \in \mu^{-1}(\xi)$ the stabilizer must be discrete:

$$\mathrm{Ann}(\mathfrak{g}_x) = \mathrm{Im}T_x\mu = \mathfrak{g}^* \quad \Leftrightarrow \quad \mathfrak{g}_x = \{0\}.$$

Now for any $\eta \in \mathcal{O}_\xi$ we have some $g \in G$ with $\eta = \mathrm{Ad}_g^*\xi$. Then the stabilizer for any point $y \in \mu^{-1}(\eta)$ is conjugate to the stabilizer of $g^{-1} \cdot y \in \mu^{-1}(\xi)$ which we know to be discrete. Therefore, the stabilizers $\mathrm{Stab}(y)$ are discrete for every $y \in \mu^{-1}(\eta)$ and so their Lie algebras are trivial in $\mathfrak{g}$. It follows that $T_y\mu$ is surjective for every $y \in \mu^{-1}(\eta)$, and hence η is also a regular value. Since η was arbitrary we conclude that the entire coadjoint orbit must consist of regular values of the moment map. □

Lemma 6.16 *The action of G_ξ on $\mu^{-1}(\xi)$ is free if and only if the action of G on $\mu^{-1}(\mathcal{O}_\xi)$ is free.*

Proof ($\Rightarrow$) Given $x \in \mu^{-1}(\mathcal{O}_\xi)$ let $\eta = \mu(x)$ and take any $g \in G$ so that $\eta = \mathrm{Ad}_g^* \xi$. If some $h \in G$ fixes x then by applying μ we see right away that h must belong to the coadjoint stabilizer of η which is necessarily conjugate to that of ξ:

$$\eta = \mu(x) = \mu(h \cdot x) = \mathrm{Ad}_h^* \eta \quad h \in G_\eta = g G_\xi g^{-1}.$$

So we may write $h = g h_0 g^{-1}$ for some $h_0 \in G_\xi$ which must fix $g^{-1} \cdot x$. Now from the equivariance of μ we see that $g^{-1} \cdot x \in \mu^{-1}(\xi)$ and because the action of G_ξ was assume to be free it follows that h_0 and therefore also h must be the identity element in G.

The other direction ($\Leftarrow$) is trivial. $\square$

So under the assumptions of Theorem 6.14, the action of G on the preimage of the coadjoint orbit defines a principal G-bundle which we denote along with the inclusion of the preimage as below:

$$i_{\mathcal{O}_\xi} : \mu^{-1}(\mathcal{O}_\xi) \hookrightarrow M \qquad p_{\mathcal{O}_\xi} : \mu^{-1}(\mathcal{O}_\xi) \to M^{\mathcal{O}_\xi}.$$

We might hope that $i_{\mathcal{O}_\xi}^* \omega$ is a basic form so that we may push it forward to a symplectic form $M^{\mathcal{O}_\xi}$; however, this is not the case since it fails to vanish on the vertical bundle. Indeed for any point $x \in \mu^{-1}(\mathcal{O}_\xi)$ and vertical vector $w \in \ker T_x p_{\mathcal{O}_\xi}$ we want to consider $\omega_x(w, v)$ for an arbitrary v tangent to $\mu^{-1}(\mathcal{O}_\xi)$ at x. The tangent vectors v and w are described as follows. Let $x \in M$ for which $\mu(x) = \eta$. Then

$$
\begin{aligned}
T_x\left(\mu^{-1}(\mathcal{O}_\xi)\right)_x &= (T_x \mu)^{-1}(T_\eta \mathcal{O}_\xi) \\
&= (T_x \mu)^{-1}\left\{ X_\eta^{\mathrm{Ad}^*} = \mathrm{ad}_X^* \eta, \quad X \in \mathfrak{g} \right\} \\
&= \left\{ v \in T_x M \mid \exists X_v \in \mathfrak{g} \text{ s.t. } T_x \mu(v) = \mathrm{ad}_{X_v}^* \mu(x) \right\} \\
\ker T_x p_{\mathcal{O}_\xi} &= T_x(G \cdot x) \\
&= \left\{ w \in T_x M \mid \exists X_w \in \mathfrak{g} \text{ s.t. } w = (X_w)_x^\# \right\}.
\end{aligned}
$$

We see then that while $\iota_w \omega_x$ need not vanish on $T_x \mu^{-1}(\mathcal{O}_\xi)$ it is controlled by the canonical 2-form $\omega^{KKS} \in \Omega^2(\mathfrak{g}^*)$. Let $\langle \cdot, \cdot \rangle$ be the canonical pairing of $T^* M$ and TM, where $\langle \alpha, X \rangle = \alpha(X)$ for $\alpha \in T_x^* M$ and $X \in T_x M$.

$$
\begin{aligned}
\omega_x(w, v) &= -\langle T_x \mu(v) \mid X_w \rangle \\
&= -\langle \mathrm{ad}_{X_v}^* \eta \mid X_w \rangle \\
&= \langle \eta \mid \mathrm{ad}_{X_v} X_w \rangle \\
&= \langle \eta \mid [X_v, X_w] \rangle \\
&= \omega_\eta^{KKS}(\mathrm{ad}_{X_v}^* \eta, \mathrm{ad}_{X_w}^* \eta) \\
&= (\mu^* \omega^{KKS})_x(v, w).
\end{aligned}
$$

Now $\mu^*\omega^{KKS}$ is G-invariant as μ is equivariant and ω^{KKS} is invariant under the coadjoint action. Therefore, the difference $\omega - \mu^*\omega^{KKS}$ is a closed invariant 2-form and we have shown that by restricting to $\mu^{-1}(\mathcal{O}_\xi)$ we obtain a basic 2-form $i^*_{\mathcal{O}_\xi}(\omega - \mu^*\omega^{KKS})$. Define $\omega^{\mathcal{O}_\xi}$ to be the pushforward of this basic 2-form to $M^{\mathcal{O}_\xi}$, so that

$$p^*_{\mathcal{O}_\xi}\omega^{\mathcal{O}_\xi} = i^*_{\mathcal{O}_\xi}(\omega - \mu^*\omega^{KKS}).$$

The pushforward $\omega^{\mathcal{O}_\xi}$ vanishes only on $(p_{\mathcal{O}_\xi})_*X$ for which $\iota_X\omega = \iota_X\mu^*\omega^{KKS}$ and this holds only for those X tangent to the G-orbits, that is, when X is a vertical vector field and hence $(p_{\mathcal{O}_\xi})_*X = 0$. Thus $\omega^{\mathcal{O}_\xi}$ is non-degenerate and therefore defines a symplectic structure $(M^{\mathcal{O}_\xi}, \omega^{\mathcal{O}_\xi})$ which we could also call a symplectic quotient. There is a natural diffeomorphism from $M^{\mathcal{O}_\xi}$ onto M^ξ, and we should hope that this is also an isomorphism of the symplectic structures. Before investigating this, we construct a third possible candidate for the symplectic quotient.

Consider $M \times \mathcal{O}_\xi$ with the skewed-product symplectic structure $\nu = \omega \oplus (-\omega^{KKS})$ so that the component-wise action of G is Hamiltonian with moment map $\Phi(x, \eta) = \mu(x) - \eta$. Then $\Phi^{-1}(0) = \{(x, \eta) \in M \times \mathcal{O}_\xi | \mu(x) = \eta\}$ is equivariantly diffeomorphic to $\mu^{-1}(\mathcal{O}_\xi)$; we refer to Φ as the shifted moment map since it is effectively shifting μ so as to vanish on $\mu^{-1}(\mathcal{O}_\xi)$. The conditions to form the symplectic quotient (M^ξ, ω^ξ) as in Theorem 6.14 are sufficient to form the symplectic quotient at the zero level of Φ as the following lemma shows.

Lemma 6.17 ("Shifting trick") *If ξ is a regular value for μ and G_ξ acts freely and properly on $\mu^{-1}(\xi)$, then zero is a regular value for Φ and G acts freely and properly on $\Phi^{-1}(0)$.*

Proof If G_ξ acts freely on $\mu^{-1}(\xi)$, then G_η must act freely on the level $\mu^{-1}(\eta)$ for any $\eta \in \mathcal{O}_\xi$ since the stabilizers are conjugate. Now for any point $p \in \Phi^{-1}(0)$ we may write $p = (x, \mu(x))$ for $\eta = \mu(x) \in \mathcal{O}_\xi$. Suppose some $g \in G$ fixes p so that

$$(g \cdot x, \mathrm{Ad}^*_g\eta) = g \cdot p = p = (x, \eta).$$

It follows from equality in the second component that $g \in G_\eta$ but then since the action of G_η on $\mu^{-1}(\eta)$ is free the equality in the second component must force g to be the identity element. We conclude that G acts freely on $\Phi^{-1}(0)$ and zero is therefore also a regular value of Φ. □

So under the hypothesis of Theorem 6.14, we have three candidates for the symplectic quotient to consider:

$$\left(M^\xi, \omega^\xi\right) \qquad \left(M^{\mathcal{O}_\xi}, \omega^{\mathcal{O}_\xi}\right) \qquad \left((M \times \mathcal{O}_\xi)^0, \nu^0\right).$$

and the following proposition ensures that these are all naturally isomorphic to one another.

Proposition 6.18 *There are canonical symplectomorphisms between the three symplectic quotients constructed above.*

Proof The natural inclusion of $\mu^{-1}(\xi)$ into $\mu^{-1}(\mathcal{O}_\xi)$ descends to a smooth map ϕ : $M^\xi \to M^{\mathcal{O}_\xi}$. The graph of the moment map provides an equivariant diffeomorphism of $\mu^{-1}(\mathcal{O}_\xi)$ onto $\Phi^{-1}(0)$. Consider the graph of μ when restricted to $\mu^{-1}(\mathcal{O}_\xi)$ and denote it by $\mathrm{Gr}(i^*_{\mathcal{O}_\xi}\mu) = i^*_{\mathcal{O}_\xi}(\mathrm{Id}_M \times \mu)$. This is necessarily an injection since it is injective in the first component and equivariant since both the inclusion and moment map are. Therefore, $\mathrm{Gr}(i^*_{\mathcal{O}_\xi}\mu)$ factors through an equivariant diffeomorphism γ_ξ onto its image which we see below is the zero level of the shifted moment map Φ:

$$\mathrm{Im}\left(\mathrm{Gr}(i^*_{\mathcal{O}_\xi}\mu)\right) = \left\{(x, \mu(x)) \in M \times \mathcal{O}_\xi \mid x \in \mu^{-1}(\mathcal{O}_\xi)\right\}$$
$$= \left\{(x, \eta) \in M \times \mathcal{O}_\xi \mid \mu(x) = \eta \in \mathcal{O}_\xi\right\}$$
$$= \Phi^{-1}(0)$$
$$\mathrm{Gr}(i^*_{\mathcal{O}_\xi}\mu) = i_0 \circ \gamma_\xi$$

An equivariant diffeomorphism between principal G-bundles γ_ξ descends to a diffeomorphism ψ on the orbit spaces. Similarly, the natural inclusion on $\mu^{-1}(\xi)$ into $\mu^{-1}(\mathcal{O}_\xi)$ descends to an injective immersion ϕ on the orbit spaces. All of this is summarized in the following commutative diagram:

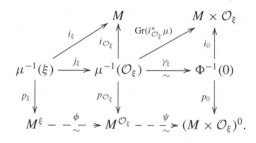

It is straightforward to verify that ϕ is surjective and hence a diffeomorphism. To see that ϕ is a symplectomorphism, we establish the following:

$$p_\xi^*\phi^*\omega^{\mathcal{O}_\xi} = j_\xi^* p_{\mathcal{O}_\xi}^*\omega^{\mathcal{O}_\xi}$$
$$= j_\xi^* i_{\mathcal{O}_\xi}^* \left(\omega - \mu^*\omega^{KKS}\right)$$
$$= i_\xi^*\omega - (\mu \circ i_\xi)^*\omega^{KKS}.$$

The first equality follows from commutativity of the diagram, the second uses the pushforward property defining $\omega^{\mathcal{O}_\xi}$ and the third is just a composition of inclusions. Moreover, the pullback $(\mu \circ i_\xi)^*\omega^{KKS}$ must vanish because μ is constant on $\mu^{-1}(\xi)$. Hence, the pullback $\phi^*\omega^{\mathcal{O}_\xi}$ satisfies the universal property defining ω^ξ and so by uniqueness they must coincide:

$$p_\xi^* \phi^* \omega^{\mathcal{O}_\xi} = i_\xi^* \omega \qquad \Rightarrow \qquad \phi^* \omega^{\mathcal{O}_\xi} = \omega^\xi.$$

We similarly verify that ψ is a symplectomorphism by showing that $\psi^* \nu^0$ satisfies the universal property defining $\omega^{\mathcal{O}_\xi}$. This is established by the following string of equalities using commutativity of the diagram, the factorization $\mathrm{Gr}(i_{\mathcal{O}_\xi}^*, \mu) = i_0 \circ \gamma_\xi$ and the definition of ν,

$$
\begin{aligned}
p_{\mathcal{O}_\xi}^* \psi^* \nu^0 &= \gamma_\xi^* p_0^* \nu^0 \\
&= \gamma_\xi^* i_0^* \nu \\
&= (i_{\mathcal{O}_\xi} \oplus \mu \circ i_{\mathcal{O}_\xi})^* (\omega \oplus -\omega^{KKS}) \\
&= i_{\mathcal{O}_\xi}^* \omega - (\mu \circ i_{\mathcal{O}_\xi})^* \omega^{KKS} \\
&= i_{\mathcal{O}_\xi}^* \left(\omega - \mu^* \omega^{KKS} \right).
\end{aligned}
$$

$\square$

Corollary 6.19 *For any two points ξ and η in the same coadjoint orbit, there is a natural isomorphism of the symplectic quotients $(M^\xi, \omega^\xi) \cong (M^\eta, \omega^\eta)$.*

Proof Both are isomorphic to $(M^{\mathcal{O}_\xi = \mathcal{O}_\eta}, \omega^{\mathcal{O}_\xi = \mathcal{O}_\eta})$. $\square$

Remark 6.20 Whenever we can form the symplectic quotient $\mu^{-1}(\xi)/G_\xi$, we may choose instead to work with the symplectic quotient at the zero level of the shifted moment map Φ. We may always assume that our symplectic quotients are formed at the zero level of the moment map, and in particular are G-orbit spaces. Unless specified otherwise, the symplectic quotients in the rest of the section are assumed to be taken at the zero level of the given moment map.

6.4   Reduction in Stages

Assume that in addition to the Hamiltonian action of a connected compact Lie group G on the connected symplectic manifold (M, ω) with moment map μ_G we have a Lie subgroup H of G. Then there is an induced Hamiltonian action of H on M with moment map given by projecting μ_G onto $\mathfrak{h}^*$. If we form the symplectic quotient at a regular level of this projected moment map, it may carry some residual symmetry from the original G-action.

Consider the natural short exact sequence associated with the inclusion $j : H \to G$ of a normal Lie subgroup H of G. Differentiating and then dualizing, this gives two more short exact sequences as below:

$$0 \longrightarrow H \overset{j}{\longrightarrow} G \overset{q}{\longrightarrow} G/H \longrightarrow 0$$

$$0 \longrightarrow \mathfrak{h} \overset{J}{\longrightarrow} \mathfrak{g} \overset{Q}{\longrightarrow} (\mathfrak{g}/\mathfrak{h}) \longrightarrow 0$$

$$0 \longleftarrow \mathfrak{h}^* \overset{J^*}{\longleftarrow} \mathfrak{g}^* \overset{Q^*}{\longleftarrow} (\mathfrak{g}/\mathfrak{h})^* \longleftarrow 0$$

The action of H on M has moment map $\mu_H = J^* \circ \mu_G$ and if we assume that zero is a regular level for μ_H on which H acts freely then we may form the symplectic quotient which we will denote by $(M^H = \mu_H^{-1}(0)/H, \omega^H)$. The kernel of J^* consists of those $\xi \in \mathfrak{g}^*$ such that for every $X \in \mathfrak{h}$ the pairing $\langle J^*\xi, X \rangle = \langle \xi, JX \rangle$ vanishes. Identifying $\mathfrak{h}$ as a subspace of $\mathfrak{g}$ via the linear map J, this means that $\ker J^* = \mathrm{Ann}(\mathfrak{h})$. Now conjugation by G preserves the normal subgroup H, and therefore the adjoint action of G preserves the Lie subalgebra $\mathfrak{h} \subset \mathfrak{g}$. If follows that G preserves the annihilator of $\mathfrak{h}$ and therefore also the zero level of μ_H

$$\mu_H^{-1}(0) = \mu_G^{-1}(\ker J^*) = \mu_G^{-1}(\mathrm{Ann}(\mathfrak{h})).$$

Normality of H implies furthermore that G preserves the H-orbits and therefore acts on the orbit space M^H in such a way that p_H is equivariant. Finally, this action on M^H clearly reduces to an action of the quotient group G/H which we will see is Hamiltonian.

Proposition 6.21 *Let M_H be the symplectic quotient with principal H-bundle denoted by $p_H : \mu_H^{-1}(0) \to M_H$ and the inclusion of the zero level by*

$$i_H : \mu_H^{-1}(0) \to M.$$

There is a natural Hamiltonian action of the quotient group G/H on M_H with moment map $\mu_{G/H}$ satisfying $Q^* \circ (p_H^* \mu_{G/H}) = i_H^* \mu_G$.

Remark 6.22 If we use Q^* to identify $(\mathfrak{g}/\mathfrak{h})^*$ with its image in $\mathfrak{g}^*$, then the relation between the moment maps becomes the same as the relation between the symplectic form on M and the reduced form ω^0,

$$p_H^* \mu_{G/H} = i_H^* \mu_G \quad \leftrightarrow \quad p_H^* \omega^\xi = i_H^* \omega.$$

Proof Having already seen that the quotient G/H acts on the orbit space M^H, it remains to verify that this action is Hamiltonian and the moment map satisfies the given relation. Because H acts trivially on $\mathrm{Ann}(\mathfrak{h})$, the equivariance of μ_G makes the restriction of μ_G to $\mu_H^{-1}(0)$ constant on H-orbits and it therefore descends to a smooth map from the orbit space $M_H \to \mathrm{Ann}(\mathfrak{h})$. Furthermore, since $\mathrm{Ann}(\mathfrak{h}) = \ker J^* = \mathrm{Im} Q^*$, this map factors uniquely through Q^*. We summarize this in the diagram below:

$$0 \longrightarrow (\mathfrak{g}/\mathfrak{h})^* \xrightarrow{\ Q^*\ } \mathfrak{g}^* \xrightarrow{\ J^*\ } \mathfrak{h}^* \longrightarrow 0$$

with $\exists ! \mu_{G/H}$ mapping up from M_H, and $i_H^* \mu_G$ mapping up to $\mathfrak{g}^*$, and $M_H \xleftarrow{\ p_H\ } \mu_H^{-1}(0)$.

It follows from equivariance of μ_G that $\mu_{G/H}$ is equivariant with respect to the action of G/H. Indeed, let $gH \in G/H$ and $H \cdot x \in M^H$; to show

$$\mu_{G/H}(gH \cdot (H \cdot x)) = \mathrm{Ad}^*_{gH} \mu_{G/H}(H \cdot x)$$

it suffices to check this equality holds when paired with any element of $\mathfrak{g}/\mathfrak{h}$ which may be written as QX for some $X \in \mathfrak{g}$. This is verified by the following string of equalities where we use the commutativity of the diagram above and the equivariance of both μ_G and Q:

$$\begin{aligned}
\langle \mu_{G/H}(gH \cdot (H \cdot x)), QX \rangle &= \langle Q^* \mu_{G/H}(p_H(g \cdot x)), X \rangle \\
&= \langle \mu_G(g \cdot x), X \rangle \\
&= \langle \mathrm{Ad}^*_g \mu_G(x), X \rangle \\
&= \langle \mathrm{Ad}^*_g Q^* \mu_{G/H}(H \cdot x), X \rangle \\
&= \langle \mathrm{Ad}^*_{gH} \mu_{G/H}(H \cdot x), QX \rangle.
\end{aligned}$$

This establishes equivariance of $\mu_{G/H}$, and it remains only to verify the moment map property. We must check that at every point $H \cdot x \in M^H$ and for every $v \in \mathfrak{g}/\mathfrak{h}$ and $w \in T_{H \cdot x} M^H$ we have

$$\iota_{v^\#} \omega^H_{H \cdot x}(w) = -\langle T_{H \cdot x} \mu_{G/H}(w), v \rangle.$$

Choose $X \in \mathfrak{g}$ and $Y \in T_x \mu_H^{-1}(0)$ so that $QX = v$ and $T_x p_H(Y) = w$. Using the fact that the projection p_H pushes forward the fundamental vector fields $(QX)^\#_x = T_x p_H(X^\#_x)$, the commutativity of the diagram, and the moment map property for μ_G we arrive at the desired equality

$$\begin{aligned}
\iota_{v^\#} \omega^H_{H \cdot x}(w) &= \omega^H_{p_H(x)}(T_x p_H(X^\#_x), T_x p_H(Y)) \\
&= (p_H^* \omega^H)_x(X^\#_x, Y) \\
&= (i_H^* \omega)_x(X^\#_x, Y) \\
&= -\langle T_x \mu_G(Y), X \rangle \\
&= -\langle Q^* T_{p_H(x)} \mu_{G/H} T_x p_H Y, X \rangle \\
&= -\langle T_{p_H(x)} \mu_{G/H}(w), v \rangle.
\end{aligned}$$

$\square$

Reduction by a normal subgroup is simplified in the case of a product group $G = G_1 \times G_2$ with $H = G_1$. In this situation, the reduced moment map is essentially the moment map for the action of the G_2 component.

Corollary 6.23 *Suppose (M, ω) carries a Hamiltonian action of $G = G_1 \times G_2$ with moment map μ and M^1 is the symplectic quotient at the zero level of the moment map μ_1 for G_1. Then the moment map μ_{G/G_1} for the action of G/G_1 on M^1 satisfies $p_{G_1}^* \mu_{G/G_1} = i_{G_1}^* \mu_2$ where p_1 and i_1 are the G_1-bundle and $\mu_1^{-1}(0)$ inclusion maps, respectively.*

Proof The splitting $G = G_1 \times G_2$ induces the identifications $\mathfrak{g} = \mathfrak{g}_1 \oplus \mathfrak{g}_2$ and $\mathfrak{g}^* = \mathfrak{g}_1^* \oplus \mathfrak{g}_2^*$ so that Q is the projection onto the second factor and Q^* is the inclusion $\xi \mapsto (0, \xi)$. Then the composition QQ^* is just the identity on $\mathfrak{g}_2^*$ and $Q\mu = \mu_2$ where $\mu = \mu_1 \oplus \mu_2$. Applying Q to the relation defining the moment map μ_{G/G_1} for the $G_2 = G/G_1$ action on M^1 produces the result

$$QQ^*(p_{G_1}^* \mu_{G/G_1}) = Qi_{G_1}^* \mu \quad \Rightarrow \quad p_{G_1}^* \mu_{G/G_1} = i_{G_1}^* \mu_2. \qquad \square$$

If G acts freely on $\mu_G^{-1}(0)$ and H acts freely on $\mu_H^{-1}(0)$, then G/H acts freely on $\mu_{G/H}^{-1}(0)$ and we can form the symplectic quotient $(M^H)^{G/H} = \mu_{G/H}^{-1}(0)/(G/H)$. This space is clearly diffeomorphic to $M^G = \mu_G^{-1}(0)/G$ and we should hope that the symplectic structures $(\omega^H)^{G/H}$ and ω^G are also isomorphic. The following theorem verifies this and effectively allows us to perform symplectic reduction in multiple stages given a tower of normal subgroups.

Theorem 6.24 (Reduction in stages) *Whenever the iterated symplectic quotient $\big((M^H)^{G/H}, (\omega^H)^{G/H}\big)$ can be formed as above, there is a natural symplectomorphism with the symplectic quotient $\big(M^G, \omega^G\big)$.*

Proof From the definition of $\mu_{G/H}$, we see that

$$\mu_G^{-1}(0) = (Q^* \circ \mu_{G/H} \circ p_H)^{-1}(0) = p_H^{-1}(\mu_{G/H}^{-1}(0))$$

and so it follows that $\mu_{G/H}^{-1}(0) = p_H(\mu_G^{-1}(0)) = \mu_G^{-1}(0)/H$. Now consider that the sub-bundle map $p_{G,G/H} : \mu_G^{-1}(0) \to \mu_{G/H}^{-1}(0)$ is equivariant with respect to the G-action in the sense that $p_{G,G/H}(g \cdot x) = gH \cdot p_{G,G/H}(x)$ and so induces a smooth map ψ on the base spaces. It is straightforward to check that ψ is a diffeomorphism. In order to see that ψ is in fact a symplectomorphism, consider the following commutative diagram incorporating the various bundle projections and inclusions,

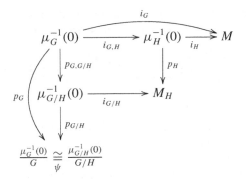

Using the universal properties defining $\omega^{G/H}$ and ω^{H} as well as the commutativity of the above diagram shows that $p_G^*(\psi^*\omega^{G/H}) = i_G^*\omega$ and so uniqueness for the reduced symplectic form ω^G on M^G implies that $\psi^*\omega^{G/H} = \omega^G$ as desired.   □

6.5 Symplectic Cutting

In this section, we introduce a construction due to Eugene Lerman called symplectic cutting [4]. This will not only provide some new examples of symplectic structures but also prove a useful theoretical tool in the following chapters. In particular, we will use symplectic cuts in the proof of the Delzant correspondence for toric manifolds. Symplectic cuts are also used to prove a nonabelian version of the convexity theorem appearing in the next chapter [5].

Suppose we have a connected symplectic manifold (M, ω) which is endowed with a Hamiltonian $U(1)$ action having moment map $H : M \to \mathbb{R}$. If we give $M \times \mathbb{C}$ the twisted product symplectic structure, then the component-wise $U(1)$ action will also be Hamiltonian with moment map given by the difference of H and half the norm squared. Any level set of this map is therefore a disjoint union of two $U(1)$-invariant subsets as described below:

$$\Phi : M \times \mathbb{C} \to \mathbb{R} \qquad \Phi : (x, w) \mapsto H(x) - \frac{1}{2}|w|^2$$

$$\Phi^{-1}(\xi) = \underbrace{H^{-1}(\xi) \times \{0\}}_{\cong H^{-1}(\xi)} \sqcup \underbrace{\bigsqcup_{r>0} H^{-1}(\xi + r) \times \left\{ \frac{1}{2}|w|^2 = r \right\}}_{\cong \{H > \xi\} \times S^1}.$$

Any regular value ξ of H is also a regular value for Φ and whenever $U(1)$ acts freely on $H^{-1}(\xi)$, so too is the action on $\Phi^{-1}(\xi)$. Indeed, the action is always free on the interior since it is free on the second component and the action is free on the boundary since it is assumed to be free on the first component. Under these conditions, we may therefore form the symplectic quotient at the level $\Phi^{-1}(\xi)$ which we denote by $(\overline{M}_{\geq\xi}, \overline{\omega}_{\geq\xi})$. Since the boundary and interior are invariant, the reduced space is diffeomorphic to the disjoint union of the orbit spaces as below:

$$\overline{M}_{\geq\xi} \cong \frac{H^{-1}(\xi)}{U(1)} \bigsqcup \frac{\{H > \xi\} \times S^1}{U(1)} \cong M^\xi \bigsqcup \{H > \xi\}.$$

Definition 6.25 (*Symplectic Cuts*) We call $(\overline{M}_{\geq\xi}, , \overline{\omega}_{\geq\xi})$ the *symplectic cut* of M (above ξ and with respect to H) since it is obtained by cutting M along the level $H^{-1}(\xi)$ and then performing symplectic reduction along the boundary to recover a symplectic manifold. An analogous construction produces the symplectic cut of M below ξ and with respect to H which we will denote by $(\overline{M}_{\leq\xi}, , \overline{\omega}_{\leq\xi})$. These two cut spaces can be glued together along the submanifolds equivalent to the symplectic quotient M^ξ to recover the original symplectic manifold (M, ω).

Proposition 6.26 *Assume now that (M, ω) carries a Hamiltonian action of some Lie group G with moment map $\mu : M \to \mathfrak{g}^*$ in addition to a $U(1)$ action with moment map H. If the action of G commutes with that of $U(1)$, then the cut spaces will also carry a Hamiltonian G-action.*

Proof Extend the action of G to the product space $M \times \mathbb{C}$ by letting it act trivially on the second component. This commutes with the $U(1)$ action on the product and so there is a well-defined Hamiltonian action of $G \times U(1)$ having moment map

$$\phi : (x, w) \mapsto \mu(x) \oplus \Phi(x, w).$$

Then our corollary on reduction by a subgroup for product groups implies the natural G-action on $\overline{M}_{\geq\xi}$ is Hamiltonian with moment map $\overline{\mu}_{\geq\xi}$ satisfying

$$p^*\overline{\mu}_{\geq\xi} = i^*\mu.$$

Under the identification of $\{H > \xi\}$ with an open subset of $M_{\geq\xi}$, the moment map $\overline{\mu}_{\geq\xi}$ agrees with the restriction of μ to $\{H > \xi\}$. On the subset isomorphic to M^ξ, the moment map $\overline{\mu}_{\geq\xi}$ agrees with μ. The map μ descends to M^ξ since it is necessarily constant on $U(1)$ orbits. $\square$

References

1. A.C. Da Silva, *Lectures on Symplectic Geometry*, vol. 1764, Lecture Notes in Mathematics (Springer, Berlin, 2001)
2. J. Lee, *Introduction to Smooth Manifolds*, GTM (Springer, Berlin, 2006)
3. N. Berline, E. Getzler, *Heat Kernels and Dirac Operators*, vol. 298, Grundlehren (Springer, Berlin, 1992)
4. E. Lerman, Symplectic cuts. Math. Res. Lett. **2**, 247–258 (1995)
5. E. Lerman, E. Meinrenken, S. Tolman, C. Woodward, Non-abelian convexity by symplectic cuts. Topology **37**, 245–259 (1998)

Chapter 7
Convexity

7.1 Introduction

In this chapter, we examine the properties of the image of the moment map for a
Hamiltonian torus action. One prototype was the Schur–Horn theorem [1, 2]: Given
a skew Hermitian matrix with prescribed eigenvalues, the diagonal entries form
the convex hull of the set of permutations of the eigenvalues. Finally, the theorem of
Atiyah and Guillemin–Sternberg incorporated all these results into a general theorem.

The layout of this chapter is as follows. In Sect. 7.2, we recall the background from
Morse theory necessary to prove the convexity theorem. In Sect. 7.3, we describe
almost periodic Hamiltonians. In Sect. 7.4, we prove the convexity theorem. In
Sect. 7.5, we describe examples and applications of the convexity theorem.

Let us begin by recalling a familiar example.

Example 7.1 Using reduction in stages, we obtained a Hamiltonian action of an
n-torus $\mathbb{T} = U(1)^{n+1}/S^1$ (where S^1 was the diagonal circle group) on the complex projective space $\mathbb{C}P^n$, and the resulting moment map is given in homogeneous
coordinates by

$$\mu : [z_0 : \cdots : z_n] \mapsto \frac{\frac{1}{2}\left(|z_0|^2, \cdots, |z_n|^2\right)}{\sum_{j=0}^{n} |z_j|^2} \qquad \mu : \mathbb{C}P^n \to \mathbb{R}^{n+1}.$$

Observe that the image of μ is the intersection of an affine hyperplane with the
positive orthant,

$$\mu(\mathbb{C}P^n) = H \cap \mathbb{R}_{\geq 0}^{n+1} \qquad H = \left\{ \xi \in \mathbb{R}^{n+1} \mid \langle \xi, (1, \cdots, 1) \rangle = \frac{1}{2} \right\}.$$

This intersection is an n-simplex the vertices of which correspond precisely to
the images of the fixed points of the action:

© The Author(s), under exclusive licence to Springer Nature Switzerland AG 2019
S. Dwivedi et al., *Hamiltonian Group Actions and Equivariant Cohomology*,
SpringerBriefs in Mathematics,
https://doi.org/10.1007/978-3-030-27227-2_7

$$\mu(\mathbb{C}P^n) = \Delta^n \qquad \mu : [0 : \cdots : z_j : \cdots : 0] \mapsto \frac{1}{2}e_j.$$

Similarly, the edges of the simplex are obtained as the image under μ of the two-dimensional submanifolds determined by setting all but two homogeneous coordinates to zero. There is a similar correspondence for higher dimensional submanifolds fixed by a subtorus mapping to the faces of Δ^n.

These properties occur more generally, and the relationship between convexity, connectedness of levels, fixed points and extremal points is the subject of this chapter. The main result we present is the classical convexity theorem for Hamiltonian torus actions due to Atiyah [3] and Guillemin–Sternberg [4]. We state the theorem below and postpone the proof until we establish some necessary preliminary results.

Theorem 7.2 (Convexity Theorem) *Let (M, ω) be a compact and connected $2d$-dimensional symplectic manifold endowed with the Hamiltonian action of an n-dimensional torus $\mathbb{T}$ along with a moment map $\mu : M \to \mathfrak{t}^*$. Then the image of μ is convex in $\mathfrak{t}^*$ and the nonempty levels $\mu^{-1}(\xi)$ are connected. Moreover, the fixed points form a finite union of connected symplectic submanifolds $C_1, \ldots, C_N$ on each of which the moment map is constant ($\mu(C_j) = c_j$) and $\mu(M)$ is the convex hull of the points $c_1, \ldots, c_N$.*

With the mention of the convex hull, let us briefly summarize some results on convex polytopes that will be useful here and in other chapters. A subset $U \subset \mathbb{R}^n$ is convex if the line segment connecting any two points in U is entirely contained within U. It is clear that $\mathbb{R}^n$ is itself convex and that the intersection of convex sets is convex. This allows us to define the convex hull of any subset $A \subset \mathbb{R}^n$ as the smallest convex set which contains A—the intersection of all convex sets containing A—and denoted Conv(A). A convex polytope is any set which is the convex hull of finitely many points. A convex polyhedron is any intersection of finitely many affine half-spaces. Polytopes are necessarily compact while polyhedra need not be. However, one can show that all compact polyhedra are polytopes and vice versa. Because of the convexity theorem, we will now refer to the image of the moment map of a Hamiltonian torus action as the associated moment polytope. Moreover, we will often take the polyhedron representation of the moment polytope.

Before proceeding towards a proof of the convexity theorem, we will show a linearized version. Consider the action of an n-dimensional torus $\mathbb{T} = U(1)^n$ acting on $\mathbb{C}^d$ with coordinates $(z_1, \ldots, z_d)$ and determined by weights

$$\alpha_1, \ldots, \alpha_d \in \mathbb{R}^n$$

so that

$$t \cdot z_j = t_1^{\alpha_j^1} \cdots t_n^{\alpha_j^n} z_j \qquad \alpha_j = (\alpha_j^1, \ldots, \alpha_j^n) \qquad t = (t_1, \ldots, t_n).$$

The moment map for this action is determined up to translation by a constant,

$$\mu_{\alpha,c}(z) = c + \frac{1}{2}\left(\sum_{j=1}^{n}\alpha_j^1|z_j|^2, \ldots, \sum_{j=1}^{n}\alpha_j^n|z_j|^2\right) = c + \sum_{j=1}^{n}|z_j|^2\alpha_j.$$

Since the $|z_j|^2$ range over the positive real numbers, this makes the image of $\mu_{\alpha,c}$ a convex cone in $\mathfrak{t}^* = \mathbb{R}^n$ generated by the vectors α_j and with vertex c.

$$\mu_{\alpha,c}(\mathbb{C}^d) = \left\{c + \sum_{j=1}^{n}r_j\alpha_j \,\middle|\, r_j \geq 0 \; j = 1,\ldots,n\right\} = \mathrm{Cone}_c(\alpha_1,\ldots,\alpha_n).$$

7.2 Digression on Morse Theory

Our main tool in the proof of Theorem 7.2 is Morse theory, and in this section we will survey some of the main ideas and results we will need. Consider a compact Riemannian manifold (M, g) and a smooth function $f : M \to \mathbb{R}$ and recall that a critical point of f is one where df vanishes. We denote the collection of all critical points of f by $\mathrm{Crit}(f)$. Assume furthermore that $\mathrm{Crit}(f)$ is a smooth submanifold of M. Then we can define the second derivative of f at any critical point as follows.

Definition 7.3 (*Hessian*) The *Hessian* of f at a critical point p is a symmetric bilinear form $D_p^2 f$ on the tangent space $T_p M$ defined for any two tangent vectors X and Y at p by

$$D_p^2 f(X, Y) = X(Y(f))\big|_p.$$

In a local coordinate system $(x_1, \ldots, x_n)$ about a critical point p of f, the Hessian of f at p is the matrix whose (j, k)-entry is $\frac{\partial^2 f}{\partial x_j \partial x_k}$ and so $D_p^2 f$ agrees with the familiar notion for functions on $\mathbb{R}^d$. The family of symmetric bilinear forms $p \mapsto D_p^2 f$ depends smoothly on $p \in C$ and we refer to this family as the *Hessian* of f denoted $D^2 f$. We are forced to restrict the definition of the Hessian to the critical submanifold of f to ensure that each $D_p^2 f$ it is in fact symmetric,

$$X(Y(f))\big|_p = Y(X(f))\big|_p - [X, Y](f)\big|_p = Y(X(f))\big|_p - \underbrace{df([X, Y])\big|_p}_{=0}.$$

At any critical point p, the Hessian defines a linear map $D_p^2 f : T_p M \to T_p^* M$ by contraction $D_p^2 f(v, -) : w \mapsto D_p^2 f(v, w)$. The kernel of this map must contain any vector tangent to $\mathrm{Crit}(f)$ at p since f is constant on the connected component containing p. We are interested in those functions which are minimally degenerate, in other words, those for which the kernel is precisely the tangent space to $\mathrm{Crit}(f)$ at every critical point. This definition was introduced by Kirwan in [5].

Definition 7.4 (*Morse–Bott functions*) If the critical set of a smooth function $f :$ $M \to \mathbb{R}$ is a closed submanifold of M and if the Hessian $D^2 f$ of f is non-degenerate in directions normal to $\mathrm{Crit}(f)$, then f is said to be a *Morse–Bott* function on M.

Remark 7.5 Note that this definition differs from that of a *Morse function* on M which requires that the Hessian be entirely non-degenerate at all critical points. This requirement forces the critical set of f to be discrete. The Morse functions we are interested in, those coming from moment maps, may well have non-discrete critical components and so must consider the more general Morse–Bott functions.

It follows that for a Morse–Bott function the Hessian induces a non-degenerate bilinear form Q_f on the normal bundle N to $\mathrm{Crit}(f)$. Non-degeneracy of Q_f implies that the normal is a direct sum of two vector bundles v^+ and v^- on which Q_f is positive and negative definite, respectively,

$$Q_{f,p}([X_p], [Y_p]) = D_p^2 f(X, Y) \qquad NZ = v^+ \oplus v^-.$$

Definition 7.6 (*Index*) The *index* of a Morse–Bott function f at a critical point p is defined to be the dimension of the negative normal bundle at p and we denote this by $\lambda_f(p) = \dim(v_p^-)$. Equivalently, the index of f at p is the number of negative eigenvalues of the associated quadratic form $Q_{f,p}$.

Proposition 7.7 *The index is constant on connected components of the critical set.*

Proof Let p be a critical point of f with C_0 the connected component of $\mathrm{Crit}(f)$ which contains it and consider any vector field X which is non-vanishing near p. Then $q \mapsto D_q^2 f(X, X)$ is a continuous function on C_0 and must be non-vanishing if the Hessian is to be non-degenerate on the normal bundle N_0 to C_0. Therefore, the sign of $D_q^2 f(X, X)$ is preserved near p as must be the dimension of v_p^-. The index is therefore locally constant and on the connected component C_0 must be constant. $\square$

Now we state some of the primary results surrounding Morse–Bott functions, the proofs of which can be found in Milnor's book [6].

Proposition 7.8 (Morse Lemma) *Let f be a Morse function on M and z a critical point of f contained in the critical submanifold $Z \subset M$. Then there are local coordinates $(x, y) = (x_1, \ldots, x_k, y_1, \ldots, y_{n-k})$ about z such that*
   (a) The critical submanifold Z is described by $y = 0$.
   (b) There is a quadratic form $q(x, y)$ which is non-degenerate in the y-variables so that $f(x, y) = f(z) + q(x, y)$.
   (c) There are finitely many connected components of $\mathrm{Crit}(f)$.

The major result we will need to borrow from Morse theory describes the homotopy types of levels of a Morse–Bott function. In the following, assume that f is a Morse–Bott function on M and for any value $c \in \mathbb{R}$ denote

$$M_c^+ = f^{-1}([c, \infty)) \qquad M_c^- = f^{-1}((-\infty, c]).$$

Theorem 7.9 *(a) If $f^{-1}(a, b)$ contains no critical points of f, then there are homotopy equivalences $f^{-1}(a) \simeq f^{-1}(b)$ and $N_a \simeq N_b$.*

(b) If $f^{-1}(a, b)$ contains one critical component Z, then there is a homotopy equivalence

$$N_b^+ \simeq N_a^+ \cup D(\nu_-(Z))$$

where $D(\nu_-(Z))$ is the disc bundle of the negative normal bundle of Z. Moreover, if Z is an isolated point then as a topological space N_b^+ is obtained by adding to N_a^+ a cell of dimension equal to the index of Z.

Corollary 7.10 *Suppose $f : M \to \mathbb{R}$ is a Morse function for which there is no critical manifold of index 1 or $n - 1$. Then*
(a) f has a unique local maximum and minimum and
(b) every level of f is either empty or connected.

The Morse–Bott functions we will consider in the rest of the chapter will always have even index. Since the dimension of M will be even as well, the hypotheses of the above corollary will always be valid.

7.3 Almost Periodic Hamiltonians

Return now to the setting of a Hamiltonian action of an n-torus $\mathbb{T}$ on a symplectic manifold (M, ω) of dimension $2d$ with moment map μ. We are in search of Morse–Bott functions on M which also tell us about the action. Since the moment map generally does not take values in $\mathbb{R}$, we must improvise and take a projection onto some one-dimensional subspace. The hope is that by choosing an appropriate subspace the projected moment map is not only a Morse–Bott function but also retains sufficient information about the action.

Lemma 7.11 *For $X \in \mathfrak{t}$, denote the subgroup generated by X by*

$$\mathbb{T}_X = \overline{Fix(\exp(\mathbb{R} \cdot X))}.$$

(The subgroup $\mathbb{T}_X$ is also a torus, as it is a closed subgroup of $\mathbb{T}$.) The fixed point set of $\mathbb{T}_X$ coincides with the critical submanifold of the almost periodic Hamiltonian μ^X

$$Fix(\mathbb{T}_X) = Crit(\mu^X).$$

Proof A point $p \in M$ is a critical point of μ^X if and only if $X^\#$ vanishes at p since for a moment map $d\mu^X = -\iota_{X^\#}\omega$. By linearity, this extends to the entire subspace generated by X,

$$d\mu_p^X = 0 \quad \Leftrightarrow \quad (\mathbb{R} \cdot X)_p^\# = 0.$$

It follows that the critical points p of μ^X correspond to those points where the $X^\#$ vanish

$$\mathrm{Crit}(\mu^X) = \mathrm{Fix}(\overline{\exp(\mathbb{R} \cdot X)}).$$ $\square$

If the one-parameter subgroup generated by X is dense in $\mathbb{T}$, the previous Lemma says that the critical submanifold of μ^X coincides with the fixed points for the entire $\mathbb{T}$ action. This is the case whenever X is chosen to have rationally independent coefficients. Such functions $\mu^X : M \to \mathbb{R}$ make good candidates to study the entire moment map μ since they retain the information of the fixed points and, as a result of the convexity theorem, the fixed points completely determine the moment polytope Δ as the convex hull of their images under μ. It will be useful to describe the behaviour exhibited by the functions μ^X without reference to an existing Hamiltonian action and so we make the following definition.

Definition 7.12 A smooth function $H \in C^\infty(M, \mathbb{R})$ is said to be an *almost periodic Hamiltonian* if the associated Hamiltonian vector field X_H generates a one-parameter group of diffeomorphisms $\{\exp(t X_H) \mid t \in \mathbb{R}\}$ the closure of which is a torus.

Since we are considering compact manifolds, we may choose a $\mathbb{T}$-invariant Riemannian metric g on M and an almost complex structure J so that

$$\omega(X, JY) = g(X, Y) \quad h(X, Y) = g(X, Y) + i\omega(X, Y),$$

where the bilinear form $h = g + i\omega$ defines a Hermitian metric. Let X_H denote the Hamiltonian vector field of H_θ so that the vanishing locus of X_H is Z. Then Z must coincide with the fixed point set of the action which we know to be a smooth submanifold of M,

$$\mathrm{Crit}(H_\theta) = V(X_H) = \mathrm{Fix}(\mathbb{T}) \subset M.$$

For any fixed point $z \in Z$, the action of $\mathbb{T}$ lifts to a linear action on the tangent space $T_z M$ which respects the Hermitian metric h (since the metric was chosen to be invariant for the original action of $\mathbb{T}$ on M). This is to say that $\mathbb{T}$ acts on $T_z M$ as a subgroup of $U(n)$ and there is a basis for $T_x M$ in which all elements of $\mathbb{T}$ are diagonal. The tangent space decomposes according to this basis into the tangent space to Z and complex subspaces V_j which are fixed by $\mathbb{T}$:

$$T_z M = T_z Z \oplus V_1 \oplus \cdots \oplus V_k \quad (w_1, \ldots, w_{n-k}, v_1, \ldots, v_k) \quad v_j = x_j + i y_j$$

In these coordinates, $\exp(X) \in U(n)$ acts on each V_j by multiplication by a complex scalar $e^{i\lambda_j}$. It follows that

$$X = \sum_{j=1}^{k} \lambda_j \left(x_j \frac{\partial}{\partial y_j} - y_j \frac{\partial}{\partial x_j} \right),$$

$$\mu^X = \frac{1}{2} \sum_{j=1}^{k} \underbrace{\lambda_j}_{\neq 0} \underbrace{(x_j^2 + y_j^2)}_{\text{pairs}} + o(|v|^2).$$

Because the λ_j's are all nonzero, this means the Hessian must be non-degenerate in the directions transverse to Z. Since this holds for arbitrary $z \in Z$ and combined with the fact that $Z = \mathrm{Crit}(H_\theta)$ is a smooth submanifold this means that H_θ is a Morse–Bott function. Notice that the λ_j's occur in pairs according to the real and imaginary parts of the complex coordinates so that the index of H_θ must be even on each critical component.

Lemma 7.13 *For any subgroup G of $\mathbb{T}$, the fixed point set $\mathrm{Fix}(G)$ is a symplectic submanifold.*

Proof Let ψ_g denote the diffeomorphism associated to any $g \in G$ and consider for any fixed point $p \in \mathrm{Fix}(G)$ the differential $(d\psi_g)_p$ which necessarily preserves the $\mathbb{T}$-invariant almost complex structure J,

$$d\psi_g(p) : T_p M \to T_p M \qquad d\psi_g(p) \circ J_p = J_p \circ d\phi_g(p).$$

Now let $\exp_p : T_p M \to M$ be the exponential mapping taken with respect to the chosen invariant metric g on M and suppose $\gamma : [0, 1] \to M$ is a geodesic with $\gamma(0) = p$ and $\dot{\gamma}(0) = v$ for some $v \in T_p M$. Then $c = \psi_g \circ \gamma$ is also a geodesic with $c(0) = \psi_g(p) = p$ and $\dot{c}(0) = (D\psi_g \cdot \dot{\gamma})(0) = D\psi_g \cdot v$ so that

$$\exp_p(D\psi_g \cdot v) = c(1) = (\psi_g \circ \gamma)(1) = \psi_g(\exp_p v).$$

The exponential map therefore provides a correspondence between the fixed point set of ψ_g in a neighbourhood of p and the fixed point set of $(d\psi_g)_p$. Thus, the fixed point set of G is the intersection of the eigenspaces with eigenvalue 1 of each $(d\psi_g)_p$ as g ranges over G

$$T_p \mathrm{Fix}(G) = \bigcap_{g \in G} \ker \left(\mathrm{Id} - d\psi_g(p) \right).$$

Each eigenspace is invariant under J_p since each $(d\psi_g)_p$ is a unitary transformation and therefore so is the intersection. We conclude that $T_p \mathrm{Fix}(G)$ is J_p-invariant for the ω-compatible almost complex structure J and so $\mathrm{Fix}(G)$ is a symplectic submanifold as claimed. $\qquad \square$

Proposition 7.14 *For all $X \in \mathfrak{t}$, the almost periodic Hamiltonian $\mu^X = \langle X, \mu \rangle$ is a Morse–Bott function with even-dimensional critical submanifolds of even index. Moreover, $\mathrm{Crit}(\mu^X)$ is a symplectic submanifold of M.*

Proof Begin by assuming that X has rationally independent coordinates so that it generates the entire torus. The critical points of μ^X will therefore coincide with the fixed point set of $\mathbb{T}$

$$\mathrm{Crit}(\mu^X) = \mathrm{Fix}(\mathbb{T}) = \bigcap_{t \in \mathbb{T}} \mathrm{Fix}(\psi_t).$$

Then, Lemma 7.13 implies C is a symplectic submanifold of M. To verify that μ^X is a Morse–Bott function, we must show that the Hessian vanishes precisely on C. To this end, consider that at any critical point p of μ^X

$$\ker D_p^2 \mu^X = \bigcap_{t \in \mathbb{T}} \ker \left(\mathrm{Id} - (d\psi_t)_p \right) \qquad \square$$

7.4 Proof of the Convexity Theorem

At last, we can give the proof of the convexity theorem following that given by Atiyah. We proceed by induction on $n = \dim(\mathbb{T})$ and for each n let us separate the statements of the theorem into the following three parts:

A_n: $\mu^{-1}(\xi)$ is either empty or connected for every $\xi \in \mathfrak{t}^*$.
B_n: $\mu(M)$ is convex.
C_n: There are finitely many connected components C_j of the fixed point set, $\mu(C_j) = c_j$ is a point and $\mu(M) = \mathrm{Conv}(c_1, \ldots, c_N)$.

To prove the connectedness of the levels of μ, it will be convenient to work with regular values. So before carrying on, we verify that there are enough regular values of μ.

Proposition 7.15 *The regular values of μ are dense in Δ.*

Proof Let C denote the union of all critical manifolds for μ^X as X ranges over $\mathfrak{g}$ and each critical manifold is the fixed point set of the action of a corresponding subtorus $\mathbb{T}_X$ and so we have established the middle inequality below:

$$M \backslash C = \left(\bigcup_{X \in \mathfrak{t}} \mathrm{Crit}(\mu^X) \right)^c = \bigcap_{X \in \mathfrak{t}} \mathrm{Fix}(\mathbb{T}_X)^c = \bigcap_{X \in \mathbb{Z}^{n+1}} \mathrm{Fix}(\mathbb{T}_X)^c.$$

The rightmost equality above can be seen as follows: each $\mathrm{Fix}(\mathbb{T}_X)$ will be the intersection of the fixed point sets for the action of circle subgroups whose product is $\mathbb{T}_X$. We may therefore consider only the intersection over $X \in \mathfrak{t}$ which generate circle subgroups $\mathbb{T}_X$, that is, the $X \in \mathfrak{t}$ which have rational components. Moreover, each such circle subgroup $\mathbb{T}_X$ can be obtained by rescaling X to lie on the integer lattice $\mathbb{Z}^{n+1} \subset \mathfrak{t}$.

Each $\text{Fix}(\mathbb{T}_X)$ is a proper closed submanifold of M and so has open and dense complement. Thus, the complement of C in M is a countable intersection of open dense sets and the Baire category theorem tells us that $M \backslash C$ must also be open and dense. An arbitrary $\xi \in \Delta$ can therefore be approximated by $\{\mu(x_j)\}$ for a sequence of points $x_j \in M \backslash C$. The image $\Delta = \mu(M)$ necessarily contains a neighbourhood of each $\mu(x_j)$, and by Sard's theorem we may find a sequence of regular values $\{\xi_{j,k}\}$ converging to $\mu(x_j)$ for every x_j. The diagonal sequence $\{\xi_{j,j}\}$ will then converge to ξ and so we conclude that regular values of μ are in fact dense in Δ. $\qquad\square$

Proof (A_n holds for all n) The statement A_1 is immediate, since for $n = 1$ the moment map $\mu : M \to \mathbb{R}$ is itself an almost periodic Hamiltonian. Now suppose that A_k holds for all Hamiltonian actions by a torus of dimension $k \le n$ and let $\mathbb{T}$ be a torus of dimension $n + 1$ acting in a Hamiltonian fashion on (M, ω) with moment map μ. We may assume that the action is effective since otherwise we may reduce to the action of some quotient of $\mathbb{T}$ and apply the induction hypothesis to conclude the result for regular levels of the reduced moment map $\tilde{\mu}$.

To apply the induction hypothesis, we need to work with the action of a subtorus of dimension not more than n. Take then the action of the subtorus of the first n-components and let $\hat{\mu}$ be the reduced moment map. With respect to the associated basis on $\mathfrak{t}^* = \mathbb{R}^{n+1}$, we may then decompose μ into $n + 1$ component functions and $\mu = (\hat{\mu}, \mu_{n+1})$. Similarly we have $\xi = (\hat{\xi}, \xi)$ for all $\xi \in \mathfrak{t}^*$. Now consider the restriction of μ_{n+1} to the level $Q = \hat{\mu}^{-1}(\hat{\xi})$,

$$Q = \hat{\mu}^{-1}(\hat{\xi}) = \bigcap_{j=1}^{n} \mu_j^{-1}(\xi_j) \qquad \mu_Q = \mu_{n+1}\big|_Q \qquad \mu^{-1}(\xi) = \mu_Q^{-1}(\xi_{n+1}).$$

It will suffice then to check that $\mu_Q^{-1}(\xi_{n+1})$ is connected and we do this by showing μ_Q satisfies the hypothesis of Lemma 7.13. A similar argument to that in the proof of Proposition 7.15 shows that the set of $\xi \in \mathfrak{t}^*$ for which $\hat{\xi}$ is a regular value for $\hat{\mu}$ is also dense in the image and so we will assume that this is the case.

Now a point $p \in Q$ is a critical point for μ_Q if and only if there are constants c_j so that $\sum_{j=1}^{n} c_j d\mu_j + d\mu_{n+1}$ vanishes at p. Then p is also a critical point for the almost periodic Hamiltonian μ^X where $X = (c_1, \ldots, c_n, 1) \in \mathfrak{t}$. Next, we show that the critical set of μ^X is transverse to Q at p, that is,

$$T_p M = T_p Q + T_p P \qquad P = \text{Crit}(\mu^X).$$

The Hamiltonian vector fields $X_j = X_{\mu_j}$ associated to the μ_j are necessarily tangent to $\text{Crit}(\mu^X)$ which we recall is a symplectic submanifold of M. Then for any coefficients λ_j the linear combination $X_\lambda = \sum_{j=1}^{n} \lambda_j X_j$ is also a section of TP so we conclude that the linear functionals

$$\sum_{j=1}^{n} \lambda_j (d\mu_j)_p = \sum_{j=1}^{n} \lambda_j (\iota_{X_j}\omega)_p = (\iota_{X_\lambda}\omega)_p \ne 0.$$

This proves that P and Q are transverse at p as claimed. Now $T_pQ \cap (T_pP)^\perp$ is the orthogonal complement of T_pP in T_pM. Then μ^X is non-degenerate on $T_pQ \cap (T_pP)^\perp$ as the orthogonal complement to T_pC in T_pM and as an almost periodic Hamiltonian it must have even index and coindex. The intersection $P \cap Q$ is a critical manifold for $\mu^X\big|_Q$ with even index and coindex. Now μ_Q is equal to μ^X up to a constant

$$\mu^X\big|_Q = \langle \left(\xi_1, \ldots, \xi_n, \mu_{n+1}\big|_Q \right), (c_1, \ldots c_n, 1) \rangle = \mu_Q + \sum_{j=1}^{n} c_j\xi_j.$$

If follows that the critical manifolds and the index of both μ^X and μ_Q agree and we may conclude the connectedness of the following level sets:

$$\mu_Q^{-1}(\xi_{n+1}) = Q \cap \mu_{n+1}^{-1}(\xi_{n+1}) = \mu^{-1}(\xi). \qquad \square$$

Note that the statement B_n for $n = 1$ is simply that $\mu(M)$ is connected and this is immediate since M is connected by assumption and μ is continuous. Now we show that the statement holds for all $n \geq 2$.

Proof (Proof that A_n implies B_{n+1}) To prove that the image is convex, it is enough to verify that the intersection of $\mu(M)$ with any straight line in $\mathbb{R}^{n+1}$ is either empty or connected. Any such straight line is an affine linear subspace and can be written as $\pi^{-1}(\eta)$ where π is the projection onto some codimension one subspace $V \subset \mathbb{R}^{n+1}$ and η is any element of V. We want to check the following set is empty or connected:

$$\mu(M) \cap \pi^{-1}(\eta) = \mu((\pi \circ \mu)^{-1}(\eta))$$

The composition $\pi \circ \mu$ describes the moment map for the action of the subtorus $S \subset \mathbb{T}$ corresponding to the subspace $V \subset \mathbb{R}^{n+1} = \mathfrak{t}^*$. The property A_n for this action of the subtorus then says that $(\pi \circ \mu)^{-1}(\eta)$ is either empty or connected and so too must be $\mu((\pi \circ \mu)^{-1}(\eta))$ since μ is continuous. Since π and η were arbitrary, this completes the proof. $\qquad \square$

Proof (Proof that B_n implies C_n) We have seen that the fixed point set of $\mathbb{T}$ coincides with the critical set for any almost periodic Hamiltonian μ^X satisfying $\mathbb{T}_X = \mathbb{T}$, and so Proposition 7.15 implies that this set has only finitely many connected components $C_1, \ldots, C_N$. The fixed point set $\mathrm{Fix}(\mathbb{T})$ is contained in the critical set $\mathrm{Crit}(\mu^X)$ for any $X \in \mathfrak{t}$ and therefore μ^X must be constant on any connected component C_j of C. Since this holds for all $X \in \mathfrak{t}$, it follows that μ is constant on each C_j and $\mu(C_j) = c_j$ is indeed a single point.

The image of μ is convex by assumption and certainly contains all of the c_j's and so also their convex hull; $\Delta = \mu(M) \supset \mathrm{Conv}(c_1, \ldots, c_N)$. To obtain the reverse inclusion, let $\xi \in \mathfrak{t}^*$ be any point not contained within the convex hull of the c_j's. As a compact convex set there is a hyperplane separating the convex hull and the point ξ;

that is, there is some $X \in \mathfrak{t}$ such that $\langle \eta, X \rangle < \langle \xi, X \rangle$ for all $\eta \in \text{Conv}(c_1, \ldots, c_N)$. Moreover, we may choose X to have rationally independent coordinates since the distance between these compact sets must be positive and such X are dense in $\mathfrak{t}$. By doing this, we ensure that $\text{Crit}(\mu^X)$ coincides with $\text{Fix}(\mathbb{T})$ and so the maximum of μ^X must occur at some $p \in C_j$. This proves the reverse inclusion since if ξ were to lie in the image of μ it would violate the following inequality:

$$\sup_{x \in M} \langle \mu(x), X \rangle = \langle \mu(p), X \rangle < \langle \xi, X \rangle. \qquad \square$$

7.5 Applications and Examples

Theorem 7.16 (Schur–Horn) *Let $\mathcal{H}$ be the set of Hermitian operators with spectrum $\lambda = \{\lambda_1, \ldots, \lambda_n\}$ and diagonal elements $a_{11}, \ldots, a_{nn}$, and let $A \in \mathcal{H}$. Then $(a_{11}, \ldots, a_{nn})$ is contained in the convex hull generated by the permutations of the eigenvalues $\text{Conv}(\lambda_{\sigma \in S_n})$. Conversely, any element of $\text{Conv}(\lambda_{\sigma \in S_n})$ is the diagonal for some Hermitian matrix with spectrum equal to λ.*

Proof The coadjoint action of $U(n)$ can be identified with conjugation on the space of skew Hermitian matrices $i\mathcal{H}$. Then the orbit O_λ containing the diagonal matrix $\text{diag}(i\lambda_1, \ldots, i\lambda_n)$ is precisely the skew Hermitian matrices with spectrum $i\lambda$. The torus subgroup $\mathbb{T} = U(1)^n$ of diagonal matrices in $U(n)$ acts on O_λ and has a moment map μ given by projecting onto the diagonal

$$\mu : A \mapsto (a_{11}, \ldots, a_{nn}).$$

We now know from the convexity theorem that the image of μ is the convex hull of $\{c_j = \mu(C_j)\}$. Now $A \in i\mathcal{H}$ is fixed by $\mathbb{T}$ if and only if it is diagonal so the fixed points are precisely the elements $\text{diag}(i\lambda_{\sigma(1)}, \ldots, i\lambda_{\sigma(n)})$. The moment polytope for the above torus action must lie in a hyperplane since the trace is constant on any O_λ

$$\langle (a_{11}, \ldots, a_{nn}), (1, \ldots, 1) \rangle = \sum_{j=1}^{n} a_{jj} = \text{Tr}(A) = \sum_{j=1}^{n} \lambda_j. \qquad \square$$

Example 7.17 Consider the case of three distinct eigenvalues having zero trace

$$\lambda_1 < \lambda_2 < \lambda_3 \qquad \lambda_1 + \lambda_2 + \lambda_3 = 0.$$

Then S_3 acts freely on the set $\{\lambda_1, \lambda_2, \lambda_3\}$ and so $\text{diag}(O_\lambda)$ is the convex hull of six distinct points which lie in the hyperplane orthogonal to $(1, 1, 1)$. This is a hexagon.

Consider the coadjoint representation of a compact Lie group G on $\mathfrak{g}^*$ and recall that the action on a particular orbit O is Hamiltonian with the inclusion $O \hookrightarrow \mathfrak{g}^*$ being the moment map. For a principal orbit, the type of O is the conjugacy class of maximal tori in G and the orbit is a homogenous space for the quotient. Choosing any representative $\xi \in O$ defines a maximal torus as its stabilizer $\mathbb{T} = G_\xi$ and a diffeomorphism

$$\varphi_\xi : \frac{G}{\mathbb{T}} \to O_\xi \qquad [g] \mapsto \mathrm{Ad}_g^* \xi.$$

We obtain a Hamiltonian torus action of $\mathbb{T}$ on O_ξ with moment map given by composing the inclusion with the projection $\mu : O_\xi \hookrightarrow \mathfrak{g}^* \to \mathfrak{t}^*$. The fixed points of this action correspond under φ_ξ to the quotient of the normalizer of $\mathbb{T}$ in G,

$$\mathrm{Fix}(\mathbb{T}) = \frac{N(\mathbb{T})}{\mathbb{T}} = W(G)$$

This is called the Weyl group of G and since $N(\mathbb{T})$ has finite index in G it is a finite group. Now $W(G)$ acts freely on O_ξ as a subgroup of $G/\mathbb{T}$ and is transitive on the fixed points of $\mathbb{T}$. Therefore, $W(G)$ acts freely and transitively on the image of these points under μ. In summary, we say that the Weyl group of G permutes the vertices of the moment polytope for the action of a maximal torus on a principal orbit.

Example 7.18 (Maximal torus in $SU(3)$) Consider the maximal torus $\mathbb{T}$ in $SU(3)$ given by diagonal matrices with entries in $U(1)$ and having determinant 1. We consider the coadjoint orbit O_λ of a point $\lambda = \mathrm{diag}(i\lambda_1, i\lambda_2, i\lambda_3)$, where $\lambda_1 + \lambda_2 + \lambda_3 = 0$, and assume furthermore that λ_i are distinct. Then indeed $\lambda \in \mathfrak{su}(3)$ and we have the Hamiltonian coadjoint action of $SU(3)$ on O_λ. By restricting to the subgroup $\mathbb{T}$, we have a Hamiltonian action with moment map $\mu : O_\lambda \hookrightarrow \mathfrak{su}(3)^* \twoheadrightarrow \mathfrak{t}^*$

Recall that we chose a compatible almost complex structure on M so that at a fixed point p the torus acts on the tangent space as a subgroup of $U(d)$ so we may choose a basis for $T_p M$ which simultaneously diagonalizes every element of $\mathbb{T}$.

$$T_p M = V_1 \oplus \cdots \oplus V_d \qquad \mathbb{T} = U(1) \times \cdots \times U(1) \subset U(d).$$

The (unitary) representation of $\mathbb{T}$ on any complex line V_j is described by its character

$$\alpha_j : \mathbb{T} \to S^1 \qquad t \cdot z = \alpha_j(t) z \quad t \in \mathbb{T} \quad z \in V_j$$

If we have a decomposition of the torus $\mathbb{T} = U(1) \times \cdots \times U(1)$ then α_j is determined by its value on the $U(1)$ components which are in turn determined by real numbers a_j^k such that

$$\alpha_j(t_k e_k) = t_k^{a_j^k}$$

Proposition 7.19 *Let z be a fixed point in M and $p = \mu(z) \in \mathfrak{t}^*$. Then there exist neighbourhoods U and V of z and p, respectively, so that*

$$\mu(U) = V \cap Cone_p(\alpha_1, \ldots, \alpha_n).$$

Proof First recall that the cone generated by the α_k's at p is the image of the moment map for the linear action of $\mathbb{T}$ with respect to the usual symplectic form

$$\mu_0 = p + \frac{1}{2}\left(\ldots \sum_{j=1}^{n} a_k^j |v_j|^2, \ldots\right) \qquad \omega_0 = \sum_{j=1}^{n} dv_j \wedge d\bar{v}_j$$

Here $\mu_0(T_z M)$ is the cone at p on $(\alpha_1, \ldots, \alpha_n)$. Using the equivariant Darboux theorem, we can describe the moment map μ in a neighbourhood of z in terms of μ_0. Since ω_z and ω_0 agree at the origin, there is a neighbourhood U_0 of zero in $T_z M$ and an equivariant map $\psi : U_0 \to T_z M$ preserving the origin and satisfying $\psi^* \omega_z = \omega_0$. We may assume that U_0 has been taken small enough so that the exponential corresponding to our invariant Riemannian metric provides a diffeomorphism with a neighbourhood U_z of z in M,

$$\exp_z\Big|_{U_0} : U_0 \xrightarrow{\sim} U_z.$$

Since the exponential is equivariant, the composition $\mu' = \mu \circ \exp_z$ provides a moment map for the linearized action of $\mathbb{T}$ on $(T_z M, \omega_z)$. It follows that $\psi^* \mu'$ is a moment map for the action of $\mathbb{T}$ with respect to ω_0, and therefore can only differ from μ_0 by a constant. We conclude that these maps are in fact equal since they agree at zero:

$$(\psi^* \mu')(0) = \mu'(\psi(0)) = \mu'(0) = \mu(\exp_z(0)) = \mu(z) = p = \mu_0(0).$$

Therefore, μ_z takes U_z to the image of μ_0 which we know to be contained in the given cone at p:

$$\mu(U_z) = \mu(\exp_z U_0) = \mu'(U_0) = \mu_0(U_0).$$

Claim: there is an open V in $\mathfrak{t}^*$ such that $\mu_0(U_0) = V \cap \mu_0(T_z M)$.

To verify this, recall that an action is said to be effective if the intersection of all stabilizers is trivial. For a Hamiltonian torus action to be effective on a symplectic manifold of dimension $2d$, we know we must have $\dim(\mathbb{T}) \le d$. When the torus has dimension exactly half that of M, we say that M along with the Hamiltonian action is a *toric* manifold. Toric manifolds are the subject of the next chapter, but we observe some of the properties of their moment polytopes now. $\qquad \qquad \Box$

Corollary 7.20 *The rank of μ at any point x in M is equal to the dimension of the face of Δ that contains $p = \mu(x)$.*

Proof Let k be the dimension of the stabilizer $\mathbb{T}_x = Stab(x)$. Consider the action of the subgroup $\mathbb{T}_x$ on M with moment map $\pi_x \circ \mu$, where π_x is the appropriate projection. We can choose a basis so that π_x is given by

$$\pi_x : (\xi_1, \ldots, \xi_d) \mapsto (\xi_1, \ldots, \xi_k)$$

Since x is a fixed point of this action by design, we can apply Proposition 7.19 to obtain a neighbourhood U of x and V of p for which

$$(\pi_x \circ \mu)(U) = V \cap \mathrm{Cone}_p(\alpha_1, \ldots, \alpha_k).$$

Suppose S is a circle subgroup of $\mathbb{T}$ generated by some $X \in \mathfrak{t}$ and let $H = \mu^X$ be the moment map for the inherited action of S on (M, ω). Assume that S acts freely on $H^{-1}(r)$ and form the symplectic cut $M^{H \leq r}$. Recall that $M^{H \leq r}$ has an open dense subset symplectomorphic to $H^{-1}((-\infty, r))$ in M, and its complement is symplectomorphic to the symplectic reduction $M^r = H^{-1}(r)/S$. The torus $\mathbb{T}$ acts in a Hamiltonian way on the cut space $M^{H \leq r}$, and the moment map on these components is μ. $\square$

References

1. I. Schur, Über eine Klasse von Mittelbildungen mit Anwendungen auf die Determinantentheorie. Sitzungsber. Berl. Math. Ges. **22**, 9–20 (1923)
2. A. Horn, Doubly stochastic matrices and the diagonal of a rotation matrix. Amer. J. Math. **76**, 620–630 (1954)
3. M.F. Atiyah, Convexity and commuting Hamiltonians. Bull. London Math. Soc. **14**, 1–15 (1982)
4. V. Guillemin, S. Sternberg, Convexity properties of the moment mapping I and II. Invent. Math. **67**, 491–513 (1982); **77**, 533–546 (1984)
5. F. Kirwan, *Cohomology of Quotients in Symplectic and Algebraic Geometry* (Princeton University Press, Princeton, 1984)
6. J. Milnor, *Morse Theory*, Annals of Mathematics Studies (Princeton University Press, Princeton, 1963)

Chapter 8
Toric Manifolds

8.1 Introduction

In this chapter, we restrict our attention to a compact connected symplectic manifold (M, ω). In the presence of a Hamiltonian torus action on M, we have seen that the geometry of the moment polytope $\Delta = \mu(M)$ is strongly related to the orbit structure of the action. We will study the case when there is as much symmetry as possible—when the torus is of largest possible dimension for the action to be effective. The main result of this chapter, due to Delzant, says that in the case of maximal symmetry the polytope completely determines the Hamiltonian $\mathbb{T}$-space, where $\mathbb{T}$ is a torus.

Toric manifolds are symplectic manifolds equipped with a Hamiltonian action of a torus whose dimension is half the dimension of the manifold. In this case, the image of the moment map is a convex polyhedron (called the Newton polyhedron). The prototype is $\mathbb{C}P^n$, complex projective space of complex dimension n, for which the Newton polyhedron is an n-simplex. The most basic example is the two-sphere $S^2 = \mathbb{C}P^1$, for which the Newton polyhedron is a closed interval.

The layout of this chapter is as follows. In Sect. 8.2, we define integrable systems. In Sect. 8.3, we define primitive polytopes. In Sect. 8.4, we describe the Delzant correspondence, which establishes a bijective correspondence between toric manifolds and certain types of polyhedra.

Since the stabilizers of any given orbit are conjugate to each other, we may associate to each orbit a conjugacy class called the type of the orbit. It is helpful to understand Hamiltonian actions by their orbit types. In particular, the following theorem allows us to define a distinguished orbit type—the principal orbits.

Theorem 8.1 *For a smooth action of Lie group G on a manifold M, there is an orbit type (H) for which $M_{(H)}$—the orbits of type (H)—form an open dense subset of M. Orbits of type (H) are said to be the principal orbits of the action.*

© The Author(s), under exclusive licence to Springer Nature Switzerland AG 2019   61
S. Dwivedi et al., *Hamiltonian Group Actions and Equivariant Cohomology*,
SpringerBriefs in Mathematics,
https://doi.org/10.1007/978-3-030-27227-2_8

If G is commutative the conjugacy class, (H) is a single subgroup $H \subset G$. Then, H fixes every point of the open dense subset $M_{(H)}$ and by continuity H must act trivially on all of M. If the action is effective, then this implies H must be the identity.

Proposition 8.2 *Consider an effective Hamiltonian action of a commutative Lie group G on (M, ω) with moment map μ. The interior of the moment polytope $\Delta = \mu(M)$ consists of regular values for μ and is nonempty. Moreover, the dimension of G is at most equal to half that of M.*

Proof Recall that the stabilizer G_x for a point $x \in M$ satisfies $\operatorname{Im} T_x \mu = \operatorname{Ann}(\mathfrak{g}_x)$ and since $\mathfrak{g}_x = \{0\}$ on principal orbits it follows that the map $d\mu_x$ is submersive. In particular, μ maps the open dense subset of principal orbits to an open dense subset of Δ.

Any orbit $G \cdot x$ is diffeomorphic to G/G_x and for a principal orbit this means $G \cdot x \cong G$. Since the orbits are isotropic submanifolds, it follows that $\dim G \leq \frac{1}{2} \dim M$. $\qquad\square$

This bounds the dimension of $\mathbb{T}$ which can act effectively and in a Hamiltonian way. So there is as much symmetry as possible when the torus has exactly half the dimension of the symplectic manifold on which it acts. This motivates the following definition.

Definition 8.3 A Hamiltonian $\mathbb{T}$-action on a compact and connected symplectic manifold (M, ω) of dimension $2n$ is said to be a *(symplectic) toric manifold* if $\mathbb{T}$ acts effectively and $\dim \mathbb{T} = n$.

Toric symplectic manifolds have a number of generalizations relaxing the smooth manifolds assumption. Notably, we can consider toric actions on orbifolds— topological spaces where a neighbourhood of every point can be identified with the quotient of an open set in a vector space by the action of a finite group. This increases the scope of symplectic reduction, allowing more general actions to be considered. For more information on orbifolds, see [1].

Example 8.4 Our construction of a $\mathbb{T}^n$ action on $\mathbb{C}P^n$ via symplectic reduction gives complex projective space the structure of a symplectic toric manifold. Indeed, $\mathbb{C}P^n$ is a compact and connected symplectic manifold with dimension $2n = 2 \dim(\mathbb{T}^n)$ and the action is indeed effective.

Example 8.5 (*Hirzebruch Surface*) Consider the symplectic manifold $\mathbb{C}P^1 \times \mathbb{C}P^2$ coming from the product of the standard symplectic structures on each of the projective spaces. Fix a positive integer k and let the torus $\mathbb{T}^2$ act as follows:

$$(t_1, t_2) \cdot ([w_1, w_2], [z_1, z_2, z_3]) = \left([t_1 w_1, w_2], [t_1^k z_1, z_2, t_2 z_3]\right).$$

The action is Hamiltonian with moment map given by the sum of moment maps for the action on each individual component

$$\mu\left([w_1, w_2], [z_1, z_2, z_3]\right) = \left(\frac{|w_1|^2}{|w_1|^2 + |w_2|^2} + k\frac{|z_1|^2}{|z_1|^2 + |z_2|^2 + |z_3|^2}, \frac{|z_3|^2}{|z_1|^2 + |z_2|^2 + |z_3|^2}\right)$$

Consider now the following subset:

$$H_k = \left\{([w_1, w_2], [z_1, z_2, z_3]) \, | \, w_1^k z_2 = w_2^k z_1\right\} \subset \mathbb{C}P^1 \times \mathbb{C}P^2.$$

Then H_k is a compact complex submanifold of complex codimension one as a level set of the homogeneous polynomial $w_1^k z_2 - w_2^k z_1$ and therefore inherits a symplectic structure. Moreover, the action of $\mathbb{T}^2$ preserves H_k; if $w_1^k z_2 = w_2^k z_1$, then

$$(t_1 w_1)^k z_2 = t_1^k w_1^k z_2 = t_1^k w_2^k z_1 = w_2^k(t_1^k z_1).$$

So the compact connected four-dimensional symplectic manifold H_k inherits an effective Hamiltonian action of the torus $\mathbb{T}^2$ and is therefore a toric manifold. We can determine the associated moment polytope by identifying the fixed points of the action and their image under μ.

$$([0, w_2], [0, 0, z_3]) \mapsto (0, 1)$$
$$([0, w_2], [0, z_2, 0]) \mapsto (0, 0)$$
$$([w_1, 0], [z_1, 0, 0]) \mapsto (k + 1, 0)$$
$$([w_1, 0], [0, 0, z_3]) \mapsto (1, 1)$$

We know that the vertices of Δ correspond to the fixed points of $\mathbb{T}$—points with orbit type $(\mathbb{T})$—and we have just seen that the interior of Δ corresponds to the principal orbits—points with orbit type $(\{1\})$. The next proposition spells out this relation in more detail by showing us that each of the orbit types is in correspondence with each of the open faces of Δ.

Proposition 8.6 *Suppose $(M, \omega, \mathbb{T}^n, \mu)$ is a $2n$-dimensional toric manifold with moment polytope Δ. For a point $x \in M$, let Δ_I be the unique face of Δ containing $\mu(x)$ in its relative interior. Then*

$$G_x = \exp(\mathfrak{h}_I) \qquad \mathfrak{h}_I = span\{v_i \, | \, i \in I\} \subset \mathfrak{t}.$$

Proof The stabilizer G_x acts effectively and with moment map $\mu_x = \pi \circ \mu$, where π is the projection onto $\mathfrak{g}_x^*$. Since x is a fixed point for this action, there is an induced unitary representation of G_x on $T_x M$ which decomposes as

$$T_x M = \underbrace{W_1 \oplus \cdots \oplus W_{n-k}}_{T_x \text{Fix}(G_x)} \oplus \underbrace{V_1 \oplus \cdots \oplus V_k}_{V}.$$

If $\alpha_1, \ldots, \alpha_k$ are the weights, then there are neighbourhoods U and V of z and $\pi(\mu(x))$, respectively, so that

$$(\pi \circ \mu)(U) = V \cap \pi(\Delta) = V \cap \mathrm{Cone}_{\pi(\mu(x))}(\alpha_1, \ldots, \alpha_k).$$

Since the action of G_x is effective, the weights α_i are linearly independent and so they generate a cone which contains only the trivial subspace. Then

$$\pi^{-1}(\mathrm{Cone}_{\pi(\eta)}(\alpha_1, \ldots, \alpha_k))$$

is an affine cone in $\mathfrak{t}^*$ with maximal affine subspace equal to $\ker \pi = \mathrm{Ann}(\mathfrak{g}_x)$. The cone $\pi^{-1}(\mathrm{Cone}_{\pi(\eta)}(\alpha_1, \ldots, \alpha_k))$ is equal to the tangent cone of Δ at $\mu(x)$. It follows that the maximal subspaces coincide. Hence $\mathrm{Ann}(\mathfrak{g}_x) = \mathrm{Ann}(\mathfrak{h}_I)$ and $\mathfrak{g}_x = \mathfrak{h}_I$. It remains only to show that the stabilizer group G_x is connected. To this end, consider the unitary representation of G_x on V with weights $\alpha_1, \ldots, \alpha_k$. We can identify G_x with a compact abelian subgroup of $U(V)$ and its identity component G_x^0 with a torus in $U(V)$. Since the action is effective, the weights form a basis for $\mathfrak{g}_x^*$ and so G_x^0 is a maximal torus in $U(V)$ as $\dim G_x^0 = \dim \mathfrak{g}_x = \dim_{\mathbb{C}} V$. Since maximal tori are by definition maximal connected abelian subgroups, this forces $G_x = G_x^0$ and we conclude that $G_x = \exp(\mathfrak{g}_x) = \exp(\mathfrak{h}_I)$. $\qquad\square$

Corollary 8.7 *For any closed face Δ_I, the preimage $\mu^{-1}(\Delta_I)$ is a symplectic submanifold of M.*

Proof A face Δ_I decomposes into a disjoint union of the open faces $\mathrm{int}(\Delta_J)$ for all the index sets J containing I and with Δ_J nonempty. From the previous proposition, we know that $\mu^{-1}(\mathrm{int}\Delta_J)$ is the set of points with stabilizer $H_J = \exp(\mathfrak{h}_J)$. On the other hand, $\mathrm{Fix}(H_I)$ is the union of all points with stabilizer contained in H_I and these are exactly the stabilizer groups of the form H_J where $J \supset I$ and $\Delta_J \neq \emptyset$. In summary, we have

$$\mu^{-1}(\Delta_I) = \bigcup_{J \supset I} \mu^{-1}(\mathrm{int}\Delta_J) = \bigcup_{J \supset I}\{G_x = H_J\} = \mathrm{Fix}(H_I).$$

Then results on fixed point sets from the previous chapter imply that $\mu^{-1}(\Delta_I)$ is a symplectic submanifold of dimension $2n - 2\dim(H_I)$. $\qquad\square$

There is a natural Hamiltonian action of the quotient $\mathbb{T}/H_I$ on $\mu^{-1}(\Delta_I)$ which is effective and since $\dim(\mathbb{T}/H_I) = n - \dim(H_I)$ this makes $\mu^{-1}(\Delta_I)$ a toric manifold. If we identify $\mathfrak{t}^*/\mathfrak{h}_I^*$ with the affine subspace parallel to $\mathrm{Ann}(\mathfrak{h}_I)$ in $\mathfrak{t}^*$, then the moment map for this action is $\mu\big|_{\mu^{-1}(\Delta_I)}$ and the moment polytope is exactly the face Δ_I.

8.2 Integrable Systems

Toric manifolds are a very special case of a more general structure referred to as integrable systems.

Proposition 8.8 *Let (M, ω, μ) be a toric manifold on dimension $2n$ with moment polytope Δ and let $U = \mu^{-1}(\mathrm{int}\Delta)$. There are coordinates (μ_i, θ_i) on U so that $\omega = \sum d\mu_i \wedge d\theta_i$.*

Proof Choose a basis for $\mathfrak{t}^*$ and so that the moment map can be written $\mu = (\mu_1, \ldots, \mu_n)$ and let $\phi_1, \ldots, \phi_n$ be any coordinate system on $\mathbb{T}$. The restriction $\mu\big|_U$ is a surjective submersion and also proper since M is assumed to be compact. Ehresmann's theorem [2] states that a smooth mapping $f : M \to N$ which is both a surjective submersion and proper is a locally trivial fibration. We may apply Ehresmann's theorem to conclude that $\mu\big|_U$ is a locally trivial fibration with typical fibre $\mu^{-1}(\xi) = \mathbb{T} \cdot x \cong \mathbb{T}$.; moreover, we can choose a global trivialization since $\mathrm{int}(\Delta)$ is convex (and thus contractible).

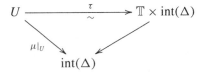

The diffeomorphism τ is equivariant with respect to the left action of $\mathbb{T}$ on itself. Now expressing ω in terms of the coordinates $\phi \circ \tau$ and $\mu \circ \tau$ on U,

$$\omega = \sum_{i,j=1}^{n} a_{ij} d\mu_i \wedge d\phi_j + \sum_{i,j=1}^{n} b_{i,j} d\mu_i \wedge d\mu_j + \sum_{i,j=1}^{n} c_{i,j} d\phi_i \wedge d\phi_j.$$

First note that all of the $c_{i,j}$ must vanish because $\omega(\partial\phi_i^\sharp, \partial\phi_i^\sharp) = c_{i,j}$ and since the fibres are Lagrangian so must be $\mathbb{T}$. The second observation we make is that $a_{ij} = \delta_{ij}$ since $-d\mu_i = \iota_{(\partial\phi_j)^\sharp}\omega = -\sum_{i=1}^{n} a_{ij} d\mu_i$. Lastly, we can see that the $b_{i,j}$ depend only on the μ_i coordinates by writing

$$\omega = \sum_{i=1}^{n} d\mu_i \wedge d\phi_i + \underbrace{\sum_{i,j=1}^{n} b_{i,j} d\mu_i \wedge d\mu_j}_{\eta}$$

and observing that η is exact (and in particular closed) as both ω and

$$\sum_{i=1}^{n} d\mu_i \wedge d\phi_i$$

are. So there are n functions $f_1, \ldots, f_n$ independent of $\mathbb{T}$ so that $\eta = d \left(\sum_{i=1}^n f_i d\mu_i \right)$. Setting θ_i as the coordinate function $\phi_i - f_i$ we are done. We have

$$\omega = \sum_{i=1}^n d\mu_i \wedge d(\theta_i + f_i) + \sum_{i=1}^n df_i \wedge d\mu_i = \sum_{i=1}^n d\mu_i \wedge d\theta_i. \qquad \square$$

8.3   Primitive Polytopes

Let Δ be the moment polytope for a toric manifold (M, ω, μ). We know that near any vertex the polytope Δ looks like the affine cone generated by the corresponding weights. This shows that there are n edges meeting any vertex and each edge is a segment of the ray emanating from the vertex in the direction of the weight. Since the weights are linearly independent, this implies that there are exactly n edges meeting any vertex of Δ.

Definition 8.9 For a vertex p of a polytope $\Delta \subset \mathbb{R}^n$, any edge that meets p is a finite segment of a ray $\{p + ru \mid r \geq 0\}$ for some $u \in \mathbb{R}^n$ unique up to a positive scalar. We say that a polytope is
  simple if for every vertex there are exactly n edges that meet p,
  rational if for every vertex the edges are generated by lattice vectors $u \in \mathbb{Z}^n$, and
  primitive if for every vertex the edges are generated by a basis for the lattice $\mathbb{Z}^n$.

It is clear that primitive implies both simple and rational; however, the converse need not hold. For example, consider the polytope $\Delta = \mathrm{conv}(\{0, e_1, e_1 + 2e_2\})$ in $\mathbb{R}^2$ with standard basis $\{e_1, e_2\}$. Then Δ is simple and rational but fails to be primitive at the origin since e_2 is not in the $\mathbb{Z}$ span of e_1 and $e_1 + 2e_2$. We have already seen that moment polytopes of toric manifolds are both simple and rational and we show now that they are in fact primitive.

Proposition 8.10 *The moment polytope for a toric symplectic manifold is a primitive polytope.*

Proof Fix a vertex η of Δ and let x be the corresponding fixed point in M and $\alpha_1, \ldots, \alpha_n \in \Lambda^*$ the weights for the unitary representation of $\mathbb{T}$ on $T_x M$. The action of $\mathbb{T}$ on $T_x M$ is determined by n characters which define $\phi \in \mathrm{Hom}(\mathbb{T}, U(1)^n)$ and for an effective action this is an isomorphism. By differentiating, we obtain an invertible linear transformation Φ, summarized below:

$$
\begin{array}{ccc}
\mathfrak{t} & \xrightarrow{\ \Phi\ } & \mathbb{R}^d \\
{\scriptstyle \exp} \big\downarrow & & \big\downarrow {\scriptstyle e^{2\pi i}} \\
\mathbb{T} & \xrightarrow[\ \phi\ ]{} & U(1)^d
\end{array}
\qquad \alpha_k = \Phi^* e_k^*
$$

Since ϕ is an isomorphism, it follows from commutativity of the diagram that $X \in \Lambda$ if and only if $\Phi(X) \in \mathbb{Z}^n$. Therefore, Φ restricts to an isomorphism of lattices, and so too does Φ^*. It follows that $\{\alpha_1, \ldots, \alpha_n\}$ is a basis for Λ^* as the image by Φ^* of the standard basis for $(\mathbb{Z}^n)^*$. There is a neighbourhood U of η in $\mathfrak{t}^*$ where $\Delta \cap V = \text{Cone}_\eta(\alpha_1, \ldots, \alpha_n) \cap V$, so the edges of Δ with common vertex η are of the form $\Delta_I = \{\eta + t\alpha_j \mid t \geq 0\} \cap \Delta$. $\qquad\square$

Recall that a convex polytope can be written as a finite intersection of half-spaces. For a convex polytope Δ in $\mathfrak{t}^*$, we write

$$\Delta = \bigcap_{i=1}^{N} \mathcal{H}^-_{v_i, h_i} \qquad \mathcal{H}^-_{v_i, h_i} = \left\{ \xi \in \mathfrak{t}^* \mid \langle \xi, v_i \rangle \leq h_i \right\} \qquad v_i \in \mathfrak{t}, \ h_i \in \mathbb{R}.$$

For a half-space $\mathcal{H}^-_{v,h}$, the pair (v, h) is unique only up to scaling by a positive constant.

Definition 8.11 A lattice vector $v \in \Lambda$ is primitive if $\frac{1}{k} v \notin \Lambda$ for any integer $k > 1$. Lattice isomorphisms preserve primitive lattice vectors. Every lattice vector can be rescaled by a positive constant to a unique primitive one. An arbitrary vector v can be rescaled to a unique primitive lattice vector if and only if the subspace $\mathbb{R}v$ intersects Λ.

Lemma 8.12 *For a primitive polytope Δ, each normal vector v can be assumed to be a primitive lattice vector. For any vertex $\Delta_I = \bigcap_{i \in I} \mathcal{H}_{v_i, h_i}$, the primitive inward normals $\{v_i\}_{i \in I}$ form a lattice basis for Λ.*

Proof For a given half-space $\mathcal{H}^-_{v_j, h_j}$, choose any vertex Δ_I contained in the facet $\Delta_{\{j\}} = \Delta \cap \mathcal{H}_{v_j, \lambda}$ with edges given by a basis $\{u_i\}_I$ for Λ^* and indexed so that $\langle u_j, v_i \rangle$ is nonzero if and only if $j = i$. Define an isomorphism $T_I^* : \mathfrak{t}^* \xrightarrow{\sim} (\mathbb{R}^n)^*$ by mapping u_j to $-e_j^*$ where $\{e_1, \ldots, e_n\}$ is the standard basis for $\mathbb{R}^n$. Then T_I^* and its adjoint T_I restrict to a lattice isomorphism and since u_i must belong to the half-space parallel to $\mathcal{H}^-_{v_i, h_i}$

$$\langle T_I^{-1} v_i, e_j^* \rangle = -\langle v_i, u_j \rangle = \begin{cases} 0 & j \neq i \\ c_I > 0 & j = i \end{cases}.$$

Therefore, $\frac{1}{c_I} v_j = T_I e_j$ is a primitive lattice vector and an outward normal for $\mathcal{H}^-_{v_j, h_j}$. For the second part, let the Δ_I be any vertex of Δ and T_I be as above so that the associated primitive outward normals obtained as the image of the standard basis for $\mathbb{Z}^n$ under the lattice isomorphism T_I and hence form a lattice basis themselves. $\qquad\square$

The converse is also true; any polytope with primitive outward normals that form a lattice basis for every vertex is a primitive polytope. The argument is similar; for any vertex, define an isomorphism a lattice isomorphism by $v_j \mapsto e_j$ and we see that the adjoint inverse image of e_j^* generates corresponding edges.

8.4   Delzant Correspondence

Theorem 8.13 *For any primitive polytope* Δ *in* $\mathfrak{t}^*$, *there exists a toric manifold with moment polytope equal to* Δ.

Recall that if a Hamiltonian $\mathbb{T}$-space (M, ω, μ) is cut below a level $h \in \mathbb{R}$ according to some Hamiltonian circle action, then there is a natural Hamiltonian $\mathbb{T}$ action on the cut space. Moreover, if the circle acts as a subgroup $\exp(\mathbb{R} \cdot X) \subset \mathbb{T}$ for some $X \in \Lambda$, then the moment polytope for the cut space is exactly the intersection of the original moment polytope with the half-space associated to X and h. We say that the new Hamiltonian $\mathbb{T}$ space is cut from the original according to the affine half-space $\mathcal{H}_{X,h}^-$.

Example 8.14 The action of $\mathbb{T}$ on itself lifts to an action on $T^*(\mathbb{T})$ which is Hamiltonian with respect to the tautological form. (In Darboux coordinates, $\{p_j, q_j\}$ the tautological form is the form $\theta = \sum_i p_i dq_i$, from which it follows that $d\theta = \omega$.) In the left trivialization $T^*(\mathbb{T}) \cong \mathbb{T} \times \mathfrak{t}^*$, the action is just multiplication in the first component and the moment map is projection onto the second. The moment polyhedron is all of $\mathfrak{t}^*$.

Think of a polytope $\Delta \subset \mathfrak{t}^*$ by successively "cutting" $\mathfrak{t}^*$ by half-spaces, and then we can try to mirror these by corresponding symplectic cuts of $T^*(\mathbb{T})$. It is clear that this process will not work for every polytope; at the very least, Δ must be rational, so that the half-spaces define circle subgroups. Assume then that Δ is a rational polytope and let the circle S_i^1 act via the subgroup of $\mathbb{T}$ generated by v_i. Let $D_0 = T^*(\mathbb{T})$ and assume that for some $i \in \{1, \ldots, N\}$ we have successfully defined a Hamiltonian $\mathbb{T}$ space D_{i-1} which carries an S_i^1 action with moment map $H_i = \mu_{i-1}^{v_i} - h_i$. If S_i^1 acts freely on the zero level of H_i, we define (D_i, ω_i, μ_i) to be the associated symplectic cut; otherwise, we terminate the process. The moment polyhedron for each D_i is obtained by cutting that of D_{i-1} by the half-space $\mathcal{H}_{v_i, h_i}^-$,

$$\mu_i(D_i) = \mu_{i-1}(D_{i-1}) \cap \mathcal{H}_{v_i, h_i}^- = \mu_0(D_0) \cap_{k=1}^i \mathcal{H}_{v_k, h_k}^- = \Delta^i.$$

If the process terminates only after $i = N$, then we denote this final stage by D_Δ. We will see shortly that up to isomorphism D_Δ does not depend on the ordering of half-spaces. By reduction in stages, the existence of D_Δ is equivalent to the existence of the symplectic quotient of $T^*(\mathbb{T}) \times \mathbb{C}^N$ for the product group $\mathbb{T}_\Delta = S_1^1 \times \cdots \times S_N^1$, and the two are isomorphic if they exist:

$$D_\Delta \cong \frac{\psi^{-1}(0)}{\mathbb{T}_\Delta} \qquad \psi = \psi_1 \oplus \cdots \oplus \psi_N \qquad \psi_i = \langle \mu_i, v_i \rangle - h_i - \frac{1}{2}|z_i|^2.$$

The existence of D_Δ therefore amounts to the action of $\mathbb{T}_\Delta$ being free on $\psi^{-1}(0)$.

Lemma 8.15 *For a primitive polytope* Δ, *the N-torus* $\mathbb{T}_\Delta$ *acts freely on the zero level of* ψ.

Proof Fix an arbitrary point $(p, z) \in \psi^{-1}(0)$ and let I denote the set of indices $i \in \{1, \ldots, N\}$ such that $z_i = 0$. For any $j \notin I$, the subgroup $S_j^1 \subset \mathbb{T}_\Delta$ acts freely at (p, z) since it acts freely on the jth component of $\mathbb{C}^N$. It remains to show that for each $i \in I$ the subgroup $S_i^1 \subset \mathbb{T}_\Delta$ acts freely at (p, z). The set I indexes the supporting hyperplanes of Δ which contain the point $\mu(p)$.

$$ i \in I \quad \leftrightarrow \quad \langle \mu(p), v_i \rangle - \lambda_i \quad \leftrightarrow \quad \mu(p) \in \mathcal{H}_{v_i, \lambda_i}. $$

Consider the homomorphism $\phi : \mathbb{T}_I \to \mathbb{T}$ which extends the embeddings of the circle groups S_i^1

$$ (t_{i_1}, \ldots, t_{i_n}) \mapsto \sum_{i \in I} t_{i_k} v_i. $$

Because the $\{v_i\}_I$ form an integer basis for the lattice, we see that the right-hand side is in the kernel of exp if and only if each t_{i_k} is an integer, which is to say that $[t]$ is the identity in $\mathbb{T}_I$. Therefore $\mathbb{T}_I$ acts on $T^*(\mathbb{T}) \times \mathbb{C}^N$ as a subgroup of $\mathbb{T}$ and is therefore free at (p, z) since $\mathbb{T}$ acts freely on the $T^*(\mathbb{T})$ component. $\qquad \square$

Now we have a Hamiltonian $\mathbb{T}$-space D_Δ with the desired moment polytope, and it remains to verify that it is in fact a toric manifold.

Proposition 8.16 *The space D_Δ is a toric manifold.*

Proof Symplectic cutting preserves dimension, so the dimension of $\mathbb{T}$ is still half that of D_Δ. Next, we see that the action must be effective, since it is so on the open dense set of D_Δ which is isomorphic to an open dense set in $T^*(\mathbb{T})$. Now we show that $\psi^{-1}(0)$ is a compact and connected set. It follows then that D_Δ is compact and connected as the orbit projection is continuous. In the left trivialization, the level $\psi^{-1}(0) \subset T^*(\mathbb{T}) \times \mathbb{C}^N$ can be written as a product of a torus and a set $Z \subset \mathfrak{t}^* \times \mathbb{C}^N$,

$$ \psi^{-1}(0) = \mathbb{T} \times Z \qquad Z = \left\{ (\xi, z) \mid \langle \xi, v_i \rangle = h_i - \frac{1}{2} |z_i|^2 \right\} $$

Since $\psi^{-1}(0)$ is closed, so is Z. For any $(\xi, z) \in Z$, it is clear that ξ must belong to Δ which is a bounded set and by Cauchy–Schwarz this provides bounds on each component of z. The set Z is therefore compact as a closed and bounded set in $\mathfrak{t}^* \times \mathbb{C}^N$ and so $\psi^{-1}(0)$ is compact as well. Finally, the convexity theorem tells us $\psi^{-1}(0)$ must be connected as a nonempty level of a moment map. $\qquad \square$

The second half of the Delzant correspondence—which we will state but not prove—says that a symplectic toric manifold is completely determined by its moment polytope.

Theorem 8.17 *If (M, ω, μ) and (M', ω', μ') are two symplectic toric manifolds with $\mu(M) = \mu'(M')$, then there is an equivariant isomorphism $\varphi : M \to M'$.*

References

1. E. Lerman, S. Tolman, Hamiltonian torus actions on symplectic orbifolds and toric varieties. Trans. Amer. Math. Soc. **349**(10), 4201–4230 (1997)
2. C. Ehresmann, Les connexions infinitésimales dans un espace fibré différentiable, in *Colloque de Topologie (Espaces Fibrés), Bruxelles*, vol. 1951 (Georges Thone, Liège/Masson et Cie, Paris, 1950), pp. 29–55

Chapter 9
Equivariant Cohomology

9.1 Introduction

Equivariant cohomology was designed to allow the study of spaces which are the quotient of a manifold M by the action of a compact group G. This can be accomplished by studying the fixed point sets of subgroups of G, notably the maximal torus T. The Cartan model replaces the study of infinite-dimensional manifolds by families of differential forms on finite-dimensional G-manifolds parametrized by an element X in the Lie algebra of G with polynomial dependence on X. A version of de Rham cohomology can be developed for the Cartan model. The localization theorem of Atiyah–Bott and Berline–Vergne describes the evaluation of such an equivariantly closed differential form on the fundamental class of the manifold.

In this chapter, we first define homotopy quotients in Sect. 9.2. We introduce the Cartan model in Sect. 9.3. We treat characteristic classes of bundles over classifying spaces in Sect. 9.4. We consider these characteristic classes in terms of the Cartan model in Sect. 9.5. We treat the equivariant first Chern class of a prequantum line bundle in Sect. 9.6. We treat Euler classes and equivariant Euler classes in Sect. 9.7 We treat the localization formula for torus actions in Sect. 9.8. Finally, we treat the abelian localization theorem of Atiyah–Bott and Berline–Vergne in Sect. 9.10.

References for this chapter are Audin [1], Sect. 5 and Berline–Getzler–Vergne [2], Sect. 7.

9.2 Homotopy Quotients

In this section, we first define classifying spaces, and then use them to define homotopy quotients.

© The Author(s), under exclusive licence to Springer Nature Switzerland AG 2019
S. Dwivedi et al., *Hamiltonian Group Actions and Equivariant Cohomology*,
SpringerBriefs in Mathematics,
https://doi.org/10.1007/978-3-030-27227-2_9

Definition 9.1 Let G be a compact Lie group. A *universal bundle* EG is a contractible space on which G acts freely. If $E_1 G$ and $E_2 G$ are two such spaces, there is a G-equivariant map between them.

Definition 9.2 The *classifying space* BG is $BG = EG/G$. For any two contractible spaces spaces $E_1 G$ and $E_2 G$ equipped with free G-actions, $E_1 G/G$ is homotopy equivalent to $E_2 G/G$.

According to Chern–Weil theory, we may obtain representatives for characteristic classes in de Rham cohomology by evaluating invariant polynomials on the curvature of a connection.

Example 9.3 S^1 acts freely on all S^{2n+1}, and these have homology only in dimensions 0 and $2n + 1$. An example of a universal space $EU(1)$ is

$$S^\infty = \{(z_1, z_2, \dots) \in \mathbb{C} \otimes \mathbb{Z} : \text{ only finitely many nonzero terms, } \sum_j |z_j|^2 = 1\}$$

$$= S^1 \cup S^3 \cup \dots$$

where $S^{2n-1} \to S^{2n+1}$ via $(z_1, \dots, z_n) \mapsto (z_1, \dots, z_n, 0)$. The space S^∞ is in fact contractible, so it is $EU(1)$.

Lemma 9.4
$$BU(1) = EU(1)/U(1) = \mathbb{C}P^\infty$$

Proposition 9.5 $H^*(BU(1)) = \mathbb{C}[x]$ *where* x *has degree* 2.

Recall
$$H^*(\mathbb{C}P^n) = \frac{\mathbb{C}[x]}{< x^{n+1} = 0 >}.$$

Suppose M is a manifold acted on by a compact Lie group (not necessarily freely). We want to find a substitute for the cohomology of the quotient space M/G. The latter object has singularities unless the action of G on M is free. We form the direct product of M by a space EG which is contractible and on which G acts freely. The resulting object is equipped with a free action of G. Because EG is contractible, $(M \times EG)/G$ has the same homotopy type as M/G.

Definition 9.6
$$H_G^*(M) = H^*(M_G)$$

where we define the homotopy quotient

$$M_G = (M \times EG)/G.$$

This diagonal quotient will often be denoted

$$M \times_G EG$$

where

$$M \times_G Y := \{(m, y) | (m, y) \sim (mg, g^{-1}y)\}$$

for M equipped with a right G action and Y equipped with a left G action.

Definition 9.7 The equivariant cohomology of a point is

$$H_G^* := H_G^*(\mathrm{pt}) = H^*(BG)$$

Proposition 9.8 *If G acts freely on M then M/G is a smooth manifold and*

$$H_G^*(M) = H^*(M/G)$$

More generally, $H_G^*(M)$ is a module over the ring $H_G^*(\mathrm{pt})$.

9.3 The Cartan Model

The Cartan model is the De Rham cohomology version of $H^*(M_G)$:

Definition 9.9
$$\Omega_G^*(M) = \left(\Omega^*(M) \otimes S(\mathbf{g}^*)\right)^G$$

where we have defined

$$S(\mathbf{g}^*) = \{f : \mathbf{g} \to \mathbb{R} : f \text{ is a polynomial}\}.$$

Here, $S(\mathbf{g}^*)$ is acted on by G through the coadjoint action of G on $\mathbf{g}^*$.

Proposition 9.10 *In the case $M = \mathrm{pt}$, we have $\Omega_G^*(\mathrm{pt}) = S(\mathbf{g}^*)^G$. Moreover,*

$$S(\mathbf{g}^*)^G = S(\mathbf{t}^*)^W$$

where the Weyl group W acts on $\mathbf{t}$.

Proof Under the adjoint action of G, every element of $\mathbf{g}$ is equivalent to an element in $\mathbf{t}$. A polynomial on $\mathbf{g}$ invariant under the action of G restricts to a polynomial on $\mathbf{t}$ invariant under the action of W. It is also true that any W-invariant polynomial f on $\mathbf{t}$ allows one to define a G-invariant polynomial $\hat{f}$ on $\mathbf{g}$ (defining $\hat{f}(Y) = g(X)$ where $X \in \mathbf{t}$ is equivalent to $Y \in \mathbf{g}$ under the adjoint action). The polynomial $\hat{f}$ is well defined because f is invariant under W. $\qquad \square$

Example 9.11 When T is abelian, we have

$$\Omega_T^*(M) = \Omega^*(M)^T \otimes S(\mathbf{t}^*)$$

since all polynomials on $\mathbf{t}$ are automatically invariant under the adjoint action, because the adjoint action of T on $\mathbf{t}$ is trivial.

Lemma 9.12

$$S(\mathbf{t}^*) = \mathbb{C}[x_1, \ldots, x_\ell]$$

where $\ell = \dim(T)$.

Let X be a formal parameter, which should be viewed as an element of the Lie algebra of a Lie group G which acts on M. The parameter X should be thought of as taking values in a vector space of dimension n if the Lie group has dimension n. We could denote X as $(X_1, \ldots, X_n)$. An element $f \in \Omega_G^*(M)$ may be thought of as a G-equivariant map $f : \mathbf{g} \to \Omega^*(M)$, where the dependence of $f(X) \in \Omega^*(M)$ on $X \in \mathbf{g}$ is polynomial. In terms of the coordinates $\{X_1, \ldots, X_n\}$, this means that we can write f as $f = \sum_{\ell \geq 0} f_\ell(X_1, \ldots, X_n)\alpha_\ell$ where $f_\ell(X_1, \ldots, X_n)$ is a polynomial in $X_1, \ldots, X_n$ and α_ℓ is a differential form of degree ℓ on M.

The grading on $\Omega_G^*(M)$ is defined by $\deg(f) = \ell + 2p$ if $X \mapsto f(X)$ is p-linear in X and $f(X) \in \Omega^\ell(M)$. We may define a differential

$$D : \Omega_G^*(M) \to \Omega_G^*(M)$$

by

$$(Df)(X) = d(f(X)) - i_{X^\#} f(X)$$

where $X^\#$ is the vector field on M generated by the action of $X \in \mathbf{g}$ and i denotes the interior product. The differential D increases the grading by 1.

Proposition 9.13 *Let G, M, $\Omega_G^*(M)$ and D be as above. Then $D \circ D = 0$.*

Proof We break down $D \circ D$ according to the degrees of differential forms. One term is $d \circ d : \Omega^k(M) \to \Omega^{k+2}(M)$ which is clearly 0. One term is $i_{X^\#} \circ i_{X^\#} : \Omega^k(M) \to \Omega^k(M)$ which is also 0 (because a differential form vanishes if it is evaluated on the same argument more than once). The last term is $d \circ i_{X^\#} + i_{X^\#} \circ d : \Omega^k(M) \to \Omega^{k+1}(M)$. This is 0 because it is the Lie derivative L_X (by Cartan's formula) and we are restricting to G-invariant forms on which the Lie derivative is 0. $\square$

Because of the previous proposition, we can make the following definitions. Define the equivariant cycles

$$Z_G^*(M) = \{\alpha \in \Omega_G^*(M) : D\alpha = 0\}$$

and the equivariant boundaries

$$B_G^*(M) = D\Omega_G^*(M).$$

Then we can define

$$H^*(\Omega_G^*(M), D) = Z_G^*(M)/B_G^*(M).$$

Theorem 9.14 (Cartan) *Let G and M be as above. The equivariant cohomology $H_G^*(M)$ of M is naturally isomorphic to the cohomology $H^*(\Omega_G^*(M), D)$ of this complex*

A good reference for Cartan's theorem is Theorem 6.1 in [3].

Proposition 9.15
$$H^*(BG) = S(\mathbf{g}^*)^G = S(\mathbf{t}^*)^W$$

*(in other words, the polynomials on **g** invariant under the adjoint action, or polynomials on **t** invariant under the Weyl group action)*

Here, the degree in $H^*(BG)$ is twice the degree as a polynomial on **g**.

Proof This follows immediately because the cohomology of BG is identified with the G-equivariant cohomology of a point, which is as above. In particular, since $D = 0$ on $\Omega_G^*(pt)$, we see that $H_G^* := H_G^*(pt) = S(\mathbf{g}^*)^G$. □

If (M, ω) is a symplectic manifold equipped with the Hamiltonian action of a compact group G, with moment map Φ, we define

$$\bar{\omega}(X) = \omega + \Phi_X \in \Omega_G^2(M).$$

The form $\bar{\omega}$ is affine linear in X. In the above grading, the equivariantly closed extension of ω has degree 2. In the next lemma, we prove that the form $\bar{\omega}$ is equivariantly closed.

Lemma 9.16
$$D\bar{\omega} = 0$$

Proof
$$(D\bar{\omega})(X) = d\omega - i_{X^\#}\omega + d\Phi_X$$

But $d\omega = 0$ and
$$i_{X^\#}\omega = d\Phi_X$$

by definition of the Hamiltonian group action. The result follows. □

We can thus define $[\bar{\omega}] \in H_G^2(M)$, where $[\alpha]$ denotes the equivalence class of α for $\alpha \in \Omega_G^*(M)$.

Example 9.17 We can view the circle as a principal circle bundle over a point p. We may equip the circle with the action of a torus T via a weight $\beta \in \text{Hom}(T, U(1))$. Thus T acts on S^1 by

$$t \in T : z \in S^1 \mapsto \beta(t)z.$$

Denote this bundle by P_β.

Example 9.18 Let $T = U(1)$ and let $\beta(t) = t^m$ for $m \in \mathbb{Z}$ be a weight. We shall denote the bundle over p (which is just a copy of $\mathbb{C}$ equipped with an action of T) by P_m.

9.4   Characteristic Classes of Bundles over $BU(1)$ and BT

Example 9.19 Let $m \in \mathbb{Z}$, and let P_m be the complex plane $\mathbb{C}$ equipped with a $U(1)$ action of weight m. The space

$$EU(1) \times_{U(1)} P_m$$

is the homotopy quotient of P_m. Explicitly this means

$$\{(z, w) \in EU(1) \times P_m\}/ \sim$$

where

$$(z, w) \sim (zu^{-1}, u^m w).$$

Every point (z, w) is equivalent to a point $(z', 1)$ by choosing $u = w^{-1/m}$ so that

$$(z, w) \sim (zw^{1/m}, 1).$$

Since there are m solutions to $u = w^{-1/m}$, any two of which differ by multiplication by a power of $e^{2\pi i/m}$, we see that

$$EU(1) \times_{U(1)} P_m = EU(1)/\mathbb{Z}_m$$

where

$$\mathbb{Z}_m = \{e^{2\pi i r/m}, r = 0, \ldots, m - 1\}.$$

We now define a connection form θ_m on $EU(1) \times_{U(1)} P_m$ (the space on the right-hand side is a complex line bundle over $BU(1)$). Note that a connection form θ on $EU(1)$ satisfies $\int_{\pi^{-1}(b)} \theta = 1$. We also require θ_m to satisfy

$$\int_{\pi_m^{-1}(b)} \theta_m = 1.$$

But since each fibre of $EU(1)$ may be written as $\{e^{i\phi}, \phi \in [0, 2\pi]\}$ and the fibre of $EU(1)/\mathbb{Z}_m \xrightarrow{\pi_m} BU(1)$ corresponds to $\{e^{i\phi}, \phi \in [0, 2\pi/m]\}$, we have $\int_{\pi_m^{-1}(b)} \theta = \frac{1}{m}$ so we need the following.

Lemma 9.20 *We have*

$$\theta_m = m\theta$$

in terms of our earlier connection form θ on $EU(1)$.

It follows that the first Chern class $c_1(P_m)$ of the principal circle bundle P_m (which is represented in Chern–Weil theory by the curvature $d\theta_m$) satisfies

Lemma 9.21

$$c_1(P_m)(X) = mX$$

where $c = c_1(EU(1) \to BU(1))$ is the generator of $H^*(BU(1))$.

Lemma 9.21 is a special case of Lemma 9.22.

Let $T \to P \to M$ be a principal T-bundle with connection

$$\theta = (\theta_1, \ldots, \theta_n) \in \Omega^1(P) \otimes \mathbf{t}.$$

Let $\beta \in \mathrm{Hom}(T, U(1))$. Choose $B \in \mathbf{t}^*$ so that $\exp(B(X)) = \beta(\exp X)$. We require also that $B(X) = 0$ for any $X \in \mathbf{t}$ for which $\exp(X) = 1$.

Form the associated principal circle bundle

$$P \times_T S^1 := \{(p, s) \in P \times S^1\}/\sim$$

where $(p, s) \sim (pt, \beta(t^{-1})s)$ for $t \in T$. Write B as

$$B = \{(b_1, \ldots, b_n)\}$$

(for $b_j \in \mathbb{Z}$). Then a connection form on $P \times_T S^1$ is

$$\sum_{j=1}^{n} b_j \theta_j = B(\theta).$$

Lemma 9.22 *If $(X_1, \ldots, X_n) \in H^2(BT)$ are the generators of $H^*(BT)$ for a torus T of rank n, then the first Chern class of the associated principal circle bundle*

$$ET \times_T (S^1)_\beta$$

specified by the weight β is

$$c_1(ET \times_T (S^1)_\beta) = \sum_{j=1}^{n} b_j X_j.$$

9.5   Characteristic Classes in Terms of the Cartan Model

Definition 9.23 A G-equivariant vector bundle over a G-manifold M is a vector bundle $V \to M$ with an action of G on the total space V covering the action of G on M.

An equivariant principal circle bundle $P \to M$ is a principal circle bundle with the action of G on the total space P covering the action of M.

Lemma 9.24 *Suppose $P \to M$ is a principal circle bundle with connection*

$$\theta \in \Omega^1(P).$$

Its first Chern class is represented in de Rham cohomology by the cohomology class of $d\theta$, denoted $c_1(P) = [d\theta]$. Note that $d\theta$ descends to a 2-form on M if and only if the bundle P is trivial; in other words, if and only if $d\theta$ is the pullback of an exact form on M.

Lemma 9.25 *If $P \xrightarrow{\pi} M$ is a G-equivariant principal $U(1)$-bundle, then its equivariant first Chern class is represented in the Cartan model by*

$$c_1^G(P) = [D\theta] = [d\theta - i_{X^\#}\theta]$$

where $X^\#$ is the vector field on P generated by $X \in \mathbf{g}$.

Proof For a collection of sections $s_\alpha : U_\alpha \subset M \to P$, $s_\alpha^* D\theta$ is closed but not exact in $\Omega_G^*(M)$. In particular, if M is a point and $P = P_\beta = U(1)$ equipped with $B \in \mathrm{Hom}(T, U(1))$, then

$$c_1(P_\beta) = [D\theta] = -\beta(X) = -i_{X^\#}\theta$$

(since $d\theta = 0$ on M). $\qquad\qquad\qquad\qquad\qquad\qquad\qquad\qquad\qquad\qquad\qquad\qquad\square$

Lemma 9.26 *If $P \to M$ is a principal $U(1)$ bundle with T action, and the T action on M is trivial (but the T action on the total space of P is not trivial), then on each fibre $\pi^{-1}(m) \cong S^1$, the T action is given by a weight $B \in \mathrm{Hom}(T, U(1))$ related to $\beta \in \mathbf{t}^*$ by*

$$B(\exp X) = \exp(\beta(X))$$

where we require $\beta(X) = 0$ if $\exp(X) = 1$. Then

$$c_1^T(P) = [D\theta] = [d\theta - \beta(X)].$$

Remark 9.27 Atiyah–Bott [4], p. 9 have a different convention on characteristic numbers. One obtains their convention from ours by replacing X by $-X$. Our convention is consistent with Berline–Getzler–Vergne [2], Sect. 7.1.

The situation of the preceding Lemma arises in the following context. If M is equipped with a G action, we apply the following Lemma where T is the maximal torus of G.

Lemma 9.28 *Let M be equipped with a T action, and let F be a component of M^T. For $\alpha \in H_T^*(M)$ and $i_F : F \to M$ the inclusion map,*

$$i_F^* \alpha \in H_T^*(F) = H^*(F) \otimes H_T^*(\mathrm{pt}) = H^*(F) \otimes \mathbb{R}[X_1, \ldots, X_n].$$

9.6 Equivariant First Chern Class of a Prequantum Line Bundle

Definition 9.29 Let (M, ω) be a symplectic manifold with a Hamiltonian action of a group G. A *prequantum line bundle with connection* is a complex line bundle $P \to M$ for which $c_1(P) = [\omega]$, equipped with a connection θ for which $d\theta = \pi^*\omega$.

Lemma 9.30 *If we impose the condition that $\mathcal{L}_{X^\#}\theta = 0$, then*

$$d i_{X^\#}\theta = -i_{X^\#}d\theta$$

$$= -i_{X^\#}\omega = -d\Phi_X.$$

It is thus natural to also impose the condition

$$i_{X^\#}\theta = -\Phi_X.$$

Thus the specification of a moment map for the group action is equivalent to specifying a lift of the action of T from M to the total space P.

Lemma 9.31 *If $F \subset M$ is a component of the fixed point set of T (the components of which will be denoted $\mathcal{F}$), then we have*

$$i_F^* \bar{\omega}(X) \in \Omega_T^2(F)$$

$$= \omega|_F + \Phi_X(F).$$

Proof For any $F \in \mathcal{F}$, the restriction $P|_F$ of the principal $U(1)$ bundle P to F is a copy of S^1 on which T acts using a weight $\exp(\beta_F) \in \mathrm{Hom}(T, U(1))$ for

$$(\beta_F) \in \mathrm{Hom}(\mathbf{t}, \mathbb{R}) = \mathbf{t}^*$$

which annihilates the kernel of the exponential map. The equivariant first Chern class of P is

$$c_1^T(P)|_F = c_1(P)|_F - \beta_F = [\omega]|_F - \beta_F.$$

Identifying the two equivariant extensions of $\omega|_F$, we see that

$$\Phi_X(F) = -(\beta_F)(X).$$

At fixed points of the action, the value of the moment map is a weight, provided the symplectic form ω satisfies $[\omega] = c_1(P)$ for some principal S^1-bundle P. This is true if and only if $[\omega] \in H^2(M, \mathbb{Z})$.   □

9.7 Euler Classes and Equivariant Euler Classes

References for this section are Roe [5], Gilkey [6] and Milnor–Stasheff [7], Appendix C.

Definition 9.32 If E is a complex vector bundle of rank m (write this as $E_\mathbb{C}$), then we may regard it as a real vector bundle of rank $2m$ (write this as $E_\mathbb{R}$).

Definition 9.33 The Euler class of E is a characteristic class $e(E)$ associated to a real vector bundle $E \to M$ of rank r, if r is the (real) dimension of M.

Definition 9.34 If $E_\mathbb{C}$ is a complex vector bundle of (complex) rank m, then $e(E_\mathbb{R}) = c_m(E_\mathbb{C})$.

Proposition 9.35 *(Euler class is multiplicative) If $E = E_1 \oplus E_2$ is the direct sum of two vector bundles, then $e(E) = e(E_1)e(E_2)$.*

Proposition 9.36 *If $E = L_1 \oplus \ldots \oplus L_m$ is the direct sum of line bundles, then $e(E) = c_1(L_1) \ldots c_1(L_m)$.*

Proposition 9.37 *If E is a complex vector bundle with a T action, and $E = \sum_j L_j$ where the L_j are complex line bundles with T action given by weights $\beta_j : T \to U(1)$, then the equivariant Euler clas of E is*

$$e^T(E) = \prod_j c_1^T(L_j)$$

which is represented in the Cartan model by

$$e^T(E)(X) = \prod_j (d\theta_j - \beta_j)(X)).$$

We can usually reduce to this situation by the *splitting principle*: see Bott–Tu [8], Sect. 21.

Example 9.38 If T acts on M and F is a component of M^T, then the normal bundle ν_F is a T-equivariant bundle over F (as the torus T acts trivially on F, but not on ν_F). Assume ν_F decomposes equivariantly as $\sum_j \nu_{F,j}$ with weights $\beta_{F,j} \in \mathbf{t}^*$.

The equivariant Euler class $e_F := e^T(\nu_F)$ is then given by

$$e_F(X) = \prod_j (c_1(\nu_{F,j}) - \beta_{F,j}).$$

References for this subsection are Berline–Getzler–Vergne [2], Sect. 7.2; Audin [1], Chap. V.6.

9.8 Localization Formula for Torus Actions

If M is a G-manifold of dimension m, then the equivariant pushforward is

$$\int_M : H_G^*(M) \to H_G^*(\mathrm{pt}).$$

Topologically, this is the pairing with the fundamental class of M. In the Cartan model, we represent an equivariant cohomology class by $\eta \in \Omega_G^*(M)$ satisfying $D\eta = 0$. Stokes' theorem implies that if M has no boundary then $\int_M D\alpha(X) = 0$ for any $\alpha \in \Omega_G^*(M)$. Thus $\int_M \eta(X)$ depends only on the class of η in the cohomology of the Cartan model. Integration over M defines a map $\int_M$ from $H_G^*(M)$ to $H_G^*(\mathrm{pt})$ which we will call the equivariant pushforward.

Remark 9.39 The integral defining the equivariant pushforward is a smooth function of X. For example, the equivariant pushforward of $e^{i\bar{\omega}}$ (where ω is the symplectic form on S^2 and $\bar{\omega}$ is its extension to an equivariantly closed form) is a constant multiple of $\sin X / X$, which is a smooth function of X. However, the terms corresponding to individual F are meromorphic functions of X which do have poles. These poles cancel in the sum over F.

The localization theorem in equivariant cohomology is stated in Sect. 9.10 and proved in Theorem 9.50. What follows is the proof of the localization theorem when M^T consists of isolated fixed points. In this case $e_F(X) = (-1)^n \prod_j \beta_{F,j}(X)$. This implies the dimension of M is even, since nontrivial irreducible representations of T have real dimension 2, and if F is a fixed point, $T_F M$ must decompose as a direct sum of nontrivial irreducible representations of T. If there were any subspaces of $T_F M$ on which T acted trivially, they would be tangent to the fixed point set M^T, but we have already assumed M^T consists of isolated fixed points.

Lemma 9.40 *Let θ be any 1-form on M for which $\theta(X^{\#}) \neq 0$ off M^T. Then on $M \setminus M^T$ we have that if $\alpha \in \Omega_T^*(M)$ and $D\alpha = 0$,*

$$\alpha = D\left(\frac{\theta\alpha}{D\theta}\right).$$

Proof (a) The differential operator D is an antiderivation (because d and $i_{X^{\#}}$ are antiderivations).

(b) So $D(\theta\alpha) = (D\theta)\alpha$ (since $D\alpha = 0$ and D is a derivation). It follows that

$$\alpha = D\left(\frac{\theta\alpha}{D\theta}\right)$$

We are using the fact that $D(f/D\theta) = Df/D\theta$.

(c) The formal expression $\frac{\theta\alpha}{D\theta}$ makes sense on $M \setminus M^T$ since $D\theta = d\theta - \theta(X^{\#})$ and $\theta(X^{\#}) \neq 0$ on $M \setminus M^T$. Then

$$\frac{1}{D\theta} = \frac{1}{-\theta(X^{\#})}\left(1 - \frac{d\theta}{\theta(X^{\#})}\right)$$

$$= \frac{-1}{\theta(X^{\#})}\sum_{r \geq 0}\left(\frac{d\theta}{\theta(X^{\#})}\right)^r$$

and $(d\theta)^r = 0$ for $2r > \dim(M)$. So the series only has a finite number of nonzero terms.   □

Lemma 9.41 *There exists a θ satisfying the hypotheses of the previous lemma.*

Proof We may construct θ on M as follows. Define θ' on $M \setminus M^T$ as follows. Choose a T-invariant metric g on M and define for $\xi \in T_m M$

$$\theta'_m(X^{\#}_m) = g(X^{\#}_m, X^{\#}_m)$$

Then $\theta'_m(X^{\#}_m) = g(X^{\#}_m, X^{\#}_m) = 1$ on $M \setminus M^T$. We choose the metric g to have this property on the complement of the fixed point set of the vector field $X^{\#}$.

In a neighbourhood of $F \in M^T$, we shall take a different choice of θ: denote it by θ''. Choose coordinates $(x_1, \ldots, x_{2n-1}, x_{2\ell})$ on $T_F M \cong \mathbb{C} \oplus \ldots \mathbb{C}$ (ℓ copies of $\mathbb{C}$) for which T acts on the jth copy of $\mathbb{C}$ (with coordinates $z_j = x_{2j-1} + ix_{2j}$) by a linear action with weight $\beta_j \in \mathbf{t}^*$, $\mathrm{Lie}(\beta_j) : \mathbf{t} \to \mathbb{R}$.

Define $\beta_j(X) = \lambda_j \in \mathbb{R}$ for a specific $X \in \mathbf{t}$ for which all the $\beta_j(X)$ are nonzero. This statement is true for almost all $X \in \mathbf{t}$.

On $\mathbb{C}^n \cong T_F M$, define

$$\theta'' = \sum_j \frac{1}{\lambda_j}(x_{2j-1}dx_{2j} - x_{2j}dx_{2j-1}).$$

The exponential map $\exp : T_F M \to M$ is T-equivariant. So

$$(x_1, \ldots, x_{2\ell})$$

become coordinates on an open neighbourhood U_F of F in M, and in these coordinates, the action of T is still given by the linear action on $\mathbb{C}^\ell$ for which the action on the jth copy of $\mathbb{C}$ is given by the weight β_j.

Using a partition of unity, construct a smooth T-invariant function

$$f : M \to [0, 1]$$

with $f = 0$ on $M \setminus U_F$. Choose an open neighbourhood $F \in V_F \subset U_F$ (for instance U_F is a ball of radius 2, and V_F is a ball of radius 1) and require $f = 1$ on V_F. Then define

$$\theta = (1 - f)\theta' + f\theta''$$

Thus

$$\theta|_{M \setminus \bigcup_{F \in M^T} U_F} = \theta'$$

and

$$\theta|_{V_F} = \theta''$$

and for appropriately chosen f, $\theta_m(X^\#) \neq 0$ when $m \notin M^T$. □

Theorem 9.42 (Stokes' theorem for the Cartan model for manifolds with boundary) *Let M be a manifold with boundary ∂M, with G action such that the action of G sends ∂M to ∂M. If $\alpha \in \Omega_G^*(M)$ then*

$$\int_M D\alpha = \int_{\partial M} \alpha.$$

Proof Decompose $\alpha = \alpha_0 + \ldots + \alpha_{\dim M}$ where α_j is a differential form of degree j (depending on X). Then $\int_M \alpha := \int_M \alpha_{\dim M}$ (by definition the other α_j contribute 0 to the integral $\int_M$). Then $(D\alpha)_{\dim M} = d\alpha_{\dim M - 1}$ (since the $i_{X^\#}$ part of the Cartan model differential reduces the degree of forms, so all terms involving $i_{X^\#}$ integrate to 0). Now apply the ordinary Stokes' theorem to $(D\alpha)_{\dim M}$. □

Let $B_\epsilon(F) \subset \exp(U_F)$ be a ball of radius ϵ around F (in the local coordinates on $\exp(U_F)$). Then

$$\int_M \alpha = \lim_{\epsilon \to 0} \int_{M \setminus \bigcup_F B_\epsilon(F)} \alpha$$

$$= \lim_{\epsilon \to 0} \int_{M \setminus \bigcup_F B_\epsilon(F)} D\left(\frac{\theta \alpha}{D\theta}\right)$$

$$= -\lim_{\epsilon \to 0} \sum_F \int_{\partial B_\epsilon(F)} \frac{\theta \alpha}{D\theta}$$

(by the usual Stokes' theorem applied to differential forms, using the fact that all terms involving $i_{X^\#}$ contribute 0). Define $\partial B_\epsilon(F) = S_\epsilon(F)$, a sphere of radius ϵ in $\mathbb{C}^\ell$.

$$S_\epsilon(F) = \{(x_1, \ldots, x_{2\ell}) : \sum_j |x_j|^2 = \epsilon^2\}.$$

Define $\phi : S^{2\ell-1} \to S_\epsilon(F)$ by $\phi(\bar{x}) = \epsilon\bar{x}$. After this rescaling, we see by considering the boundary term from the previous Lemma that the fixed point F contributes the inverse of the equivariant Euler class of the tangent space at F.   □

9.9   Equivariant Characteristic Classes

Define the pushforward map $\pi_* : \Omega_G^*(M) \to \Omega_G^*(\text{pt})$ by $\pi_*(\eta)(X) = \int_M \eta(X)$ (where $\pi : M \to$ point is projection to one point).

Stokes' theorem for equivariant cohomology in the Cartan model tells us that if M is a G-manifold with boundary and $G : \partial M \to \partial M$ (where the action of G on ∂M is locally free) and $\eta \in \Omega_G^*(M)$, then

$$\int_M (D\eta)(X) = \int_{\partial M} \eta(X).$$

It follows that the pushforward map π_* induces a map $H_G^*(M) \to H_G^*(\text{pt})$.

Definition 9.43 Suppose E is a (complex) vector bundle on a manifold M equipped with a Hamiltonian action of a group G which lifts the action of G on M. The equivariant Chern classes $c_r^G(E)$ are given by

$$c_r^G(E) = c_r(E \times_G EG \to M \times_G EG).$$

Likewise, the *equivariant Euler class* of E is given by

$$e^G(E) = e(E \times_G EG \to M \times_G EG).$$

Example 9.44 (*Equivariant characteristic classes in the Cartan model*) Suppose E is a complex vector bundle of rank N on a manifold M equipped with the action of a group G. Let ∇ be a connection on E compatible with the action of G. Define the *moment* of E, $\tilde{\mu} \in \text{End}\,E \otimes \mathbf{g}^*$ (see [2], Sect. 7.1) as follows:

$$\mathcal{L}_{X^\#}s - \nabla_{X^\#}s = \tilde{\mu}(X)s \qquad (9.1)$$

for $s \in \Gamma(E)$ (where $X \in \mathbf{g}$ and $X^\#$ is the fundamental vector field on M associated to X). Notice that the action of G on the total space of E permits us to define the Lie derivative $\mathcal{L}_{X^\#}s$ of a section $s \in \Gamma(E)$, and that the formula (9.1) defines $\tilde{\mu}$ as a zeroth-order operator (in other words, a section of $\text{End}(E)$ depending linearly on $X \in \mathbf{g}$).

We find that the representatives in the Cartan model of $c_r^G(E)$ are given by

$$c_r^G(E) = [\tau_r(F_\nabla + \tilde{\mu}(X))]$$

where $F_\nabla \in \Gamma(\mathrm{End}\,E \otimes \Omega^2(M))$ is the curvature of ∇ and τ_r is the elementary symmetric polynomial of degree r on $\mathbf{u}(N)$ giving rise to the rth Chern class c_r.

Remark 9.45 If M is symplectic and E is a complex line bundle $\mathcal{L}$ whose first Chern class is the De Rham cohomology class of the symplectic form, then the moment defined in Example 9.44 reduces to the symplectic moment map for the action of G.

Example 9.46 Suppose E is a complex line bundle over M equipped with an action of a torus T compatible with the action of T on M, and denote by F the components of the fixed point set of T over M. Suppose a torus T acts on the fibres of $E|_F$ with weight $\beta_F \in \mathbf{t}^*$. In other words, if $X \in \mathbf{t}$, then the action of $\exp(X) \in T$ sends $z \in E|_F$ to $e^{i\beta_F(X)}z$. Then in this notation, the restriction of the equivariant Euler class of E to F is given by

$$e^T(E)|_F = c_1(E) + \beta_F(X).$$

Example 9.47 If G acts on a manifold M, bundles associated to M (for example tangent and cotangent bundles) naturally acquire a compatible action of G.

Example 9.48 Suppose a torus T acts on M and let F be a component of the fixed point set. (Notice that each F is a manifold, since the action of T on the tangent space $T_f M$ at any $f \in F$ can be linearized and the linearization gives charts for F as a manifold.) Let ν_F be the normal bundle to F in M; then T acts on ν_F. Without loss of generality (as mentioned above, we are using the splitting principle: see for instance Bott and Tu [8]), we may assume that ν_F decomposes T-equivariantly as a direct sum of line bundles $\nu_{F,j}$ on each of which T acts with weight $\beta_{F,j} \in \mathbf{t}^*$. Thus one observes that the equivariant Euler class of ν_F is

$$e_F(X) = \prod_j (c_1(\nu_{F,j}) + \beta_{F,j}(X)).$$

Notice that $\beta_{F,j} \neq 0$ for any j, since otherwise $\nu_{F,j}$ would be tangent to the fixed point set rather than normal to it. We may thus define

$$e_F^0(X) = \prod_j \beta_{F,j}(X)$$

and we have

$$e_F(X) = e_F^0(X) \prod_j (1 + \frac{c_1(\nu_{F,j})}{\beta_{F,j}(X)}).$$

Since $c_1(\nu_{F,j})/\beta_{F,j}(X)$ is nilpotent (recall a class U is said to be nilpotent if there is a positive integer N for which $U^N = 0$), we find that we may define the inverse of $e_F(X)$ by

$$\frac{1}{e_F(X)} = \frac{1}{e_F^0(X)} \sum_{r=0}^{\infty} (-1)^r \left(\frac{c_1(\nu_{F,j})}{\beta_{F,j}(X)^r}\right);$$

only a finite number of terms contribute to this sum.

Example 9.49 $U(1)$ **actions with isolated fixed points**.

Suppose the action of $T \in U(1)$ on M has isolated fixed points. Suppose the normal bundle $\nu_F = T_F M$ at each fixed point F decomposes as a direct sum $\nu_F \cong \oplus_{j=1}^N \nu_{F,j}$ where each $\nu_{F,j} \cong \mathbb{C}$ and M acts with multiplicity $\mu_{F,j}$ on $\nu_{F,j}$ (for $0 \neq \mu_{F,j} \in \mathbb{Z}$): in other words

$$t \in U(1) : z_j \in \nu_{F,j} \mapsto t^{\mu_{F,j}} z_j.$$

We then find that the equivariant Euler class is

$$e_F(X) = \left(\prod_j \mu_{F,j}\right) X^N.$$

9.10   The Localization Theorem in Equivariant Cohomology

A very important localization formula for equivariant cohomology with respect to torus actions is given by the following theorem.

Theorem 9.50 (Berline–Vergne [9]; Atiyah–Bott [4]) *Let T be a torus acting on a manifold M, and let $\mathcal{F}$ index the components F of the fixed point set M^T of the action of T on M. Let $\eta \in H_T^*(M)$. Then*

$$\int_M \eta(X) = \sum_{F \in \mathcal{F}} \int_F \frac{\eta(X)}{e_F(X)}.$$

Proof (Berline–Vergne [9]) Let us assume $T = U(1)$ for simplicity. Define $M_\epsilon = M \setminus \coprod_F U_\epsilon^F$ where U_ϵ^F is an ϵ-neighbourhood (in a suitable equivariant metric) of the component F of the fixed point set M^T. On M_ϵ, T acts locally freely, so we may choose a connection θ on M_ϵ viewed as the total space of a principal (orbifold) $U(1)$ bundle (in other words, θ is a 1-form on M_ϵ for which $\theta(V) = 1$ where V is the vector field generating the S^1 action). Now for every equivariant form $\eta \in \Omega_T^*(M)$ for which $D\eta = 0$, we have that

$$\eta = D\left(\frac{\theta\eta}{d\theta - X}\right).$$

Applying the equivariant version of Stokes' theorem (Theorem 9.42), we see that

$$\int_M \eta(X) = \lim_{\epsilon \to 0} \int_{M_\epsilon} \eta(X) = \sum_F \lim_{\epsilon \to 0} \int_{\partial U_\epsilon^F} \frac{\theta\eta(X)}{d\theta - X}.$$

It can be shown (see [9] or Sect. 7.2 of [2]) that as $\epsilon \to 0$, $\int_{\partial U_\epsilon^F} \frac{\theta\eta(X)}{d\theta - X}$ tends to $\int_F \frac{\eta(X)}{e_F(X)}$.
□

Proof (Atiyah–Bott [4]) We work with the functorial properties of the pushforward (in equivariant cohomology) under the map i_F including F in M. We see that $i_F^*(i_F)_* = e_F$ is multiplication by the equivariant Euler class e_F of the normal bundle to F. Furthermore, one may show ([10], Sect. 6, Proposition 8) that the map

$$\sum_F i_F^* : H_T^*(M) \mapsto \oplus_{F \in \mathcal{F}} H^*(F) \otimes H_T^*(\text{pt})$$

is *injective*. Thus we see that each class $\eta \in H_T^*(M)$ satisfies

$$\eta = \sum_{F \in \mathcal{F}} (i_F)_* \frac{1}{e_F} i_F^* \eta \qquad (9.2)$$

(by applying i_F^* to both sides of the equation). Now $\int_M \eta = \pi_* \eta$ (where the map $\pi : M \to \text{pt}$ and $\pi_* : H_T^*(M) \to H_T^*$ is the pushforward in equivariant cohomology). The result now follows by applying π_* to both sides of (9.2) (since $\pi_* \circ (i_F)_* = (\pi_F)_* = \int_F$). □

Proof (Bismut [11]; Witten [12], 2.2.2) For more details, there is an excellent presentation of this material in the book [13]. Let $\lambda \in \Omega^1(M)$ be such that $\iota_{X^\#}\lambda = 0$ if and only if $X^\# = 0$: for instance, we may choose $\lambda(Y) = g(X^\#, Y)$ for any tangent vector Y (where g is any G-invariant metric on M). We observe that if $D\eta = 0$ then $\int_M \eta(X) = \int_M \eta(X)e^{tD\lambda}$ for any $t \in \mathbb{R}$. Now $D\lambda = d\lambda - g(X^\#, X^\#)$, so

$$\int_M \eta(X)e^{tD\lambda} = \int_M \eta(X)e^{-tg(X^\#,X^\#)} \sum_{m \geq 0} t^m (d\lambda)^m / m!.$$

Taking the limit as $t \to \infty$, we see that the integral reduces to contributions from points where $X^\# = 0$ (i.e., from the components F of the fixed point set of T). A careful computation yields Theorem 9.50.[1] □

---

[1] The technique used in this proof—introducing a parameter t, showing independence of t by a cohomological argument and showing localization as t tends to some limit—is by now universal in geometry and physics. Two of the original examples were Witten's treatment of Morse theory in [14] and the heat equation proof of the index theorem [15].

References

1. M. Audin, *Torus Actions on Symplectic Manifolds*. Progress in Mathematics, vol. 93 (Birkhäuser, 2004)
2. N. Berline, E. Getzler, M. Vergne, *Heat Kernels and Dirac Operators*. Grundlehren, vol. 298 (Springer, 1992)
3. E. Meinrenken, Equivariant cohomology and the Cartan model, *Encyclopedia of Mathematical Physics* (Elsevier, 2006)
4. M.F. Atiyah, R. Bott, The moment map and equivariant cohomology. Topology **23**, 1–28 (1984)
5. J. Roe, *Elliptic Operators, Topology and Asymptotic Methods* (Chapman and Hall, 1988)
6. P. Gilkey, *Invariance Theory, the Heat Equation and the Atiyah–Singer Index Theorem* (Publish or Perish, 1996)
7. J. Milnor, J. Stasheff, *Characteristic Classes* (Princeton University Press, Annals of Mathematics Studies, 1974)
8. R. Bott, L. Tu, *Differential Forms in Algebraic Topology*. Graduate Texts in Mathematics, vol. 82 (Springer, 1982)
9. N. Berline, M. Vergne, Zéros d'un champ de vecteurs et classes caractéristiques équivariantes. Duke Math. J. **50**, 539–549 (1983)
10. F. Kirwan, The cohomology rings of moduli spaces of vector bundles over Riemann surfaces. J. Amer. Math. Soc. **5**, 853–906 (1992)
11. J.-M. Bismut, Localization formulas, superconnections, and the index theorem for families. Commun. Math. Phys. **103**, 127–166 (1986)
12. E. Witten, Two dimensional gauge theories revisited. J. Geom. Phys. **9**, 303–368 (1992)
13. W. Zhang, *Lectures on Chern–Weil Theory and Witten Deformations* (World Scientific, 2001)
14. E. Witten, Supersymmetry and Morse theory. J. Differ. Geom. **17**, 661–692 (1982)
15. M.F. Atiyah, R. Bott, V. Patodi, On the heat equation and the index theorem. Invent. Math. **19**, 279–330 (1973)

Chapter 10
The Duistermaat–Heckman Theorem

10.1 Introduction

There are two formulations of the Duistermaat–Heckman theorem, which is the main result of Heckman's PhD thesis and is presented in [1]. The first (which comes from the original article [1]) describes how the Liouville measure of a symplectic quotient varies. The second describes an oscillatory integral over a symplectic manifold equipped with a Hamiltonian group action and can be characterized by the slogan "Stationary phase is exact".

The layout of this chapter is as follows. Section 10.2 describes the normal form theorem. Section 10.3 states the Duistermaat–Heckman theorem. Section 10.4 describes the pushforward of the Liouville measure. Section 10.6 defines the Kirwan map. Section 10.8 outlines the residue formula. Section 10.9 defines the residue formula by induction.

If (M, ω) is a compact symplectic manifold equipped with a Hamiltonian torus action with moment map, $\Phi : M \to \mathbf{t}^*$ and η^0 is a regular value in $\mathbf{t}^*$ so $M_{\eta^0} = \Phi^{-1}(\eta^0)/T$ (the symplectic quotient at η^0) is a smooth manifold or at worst an orbifold, we might ask how M_η varies as η varies in a neighbourhood of η^0 consisting of regular values of Φ.

Proposition 10.1 *The critical values of Φ are of the form $\Phi(M^{T'})$ where $M^{T'}$ is the fixed point set of a one-parameter subgroup $T' \cong U(1)$ of T.*

For topological reasons, only finitely many such subgroups will appear. The images of $\Phi(A)$, where A is a component of M^T, are subsets of intersections of $\Phi(M)$ with hypersurfaces. These hypersurfaces are normal to the vectors u ($u \in \mathbf{t}$) which generates the one-parameter subgroups T'.

Example 10.2 If $K = SU(3)$ and T is its maximal torus, the coadjoint orbit O_λ (for generic λ) is a Hamiltonian T space. The fixed point set of the T action is the

© The Author(s), under exclusive licence to Springer Nature Switzerland AG 2019   89
S. Dwivedi et al., *Hamiltonian Group Actions and Equivariant Cohomology*,
SpringerBriefs in Mathematics,
https://doi.org/10.1007/978-3-030-27227-2_10

collection of points $w\lambda$ where $w \in W$. The moment map image $\Phi_T(O_\lambda)$ is a hexagon, the convex hull of $\{w\lambda\}$. See [2], Figs. 4.3 and 4.7.

We shall see

Theorem 10.3 (Duistermaat–Heckman [1]) *If η^0 is a regular value of Φ, then for η in a sufficiently small neighbourhood U of η^0, $M_\eta \cong M_{\eta^0}$ (the two are diffeomorphic). However, M_η is not symplectically diffeomorphic to M_{η^0}. In fact, identifying the symplectic forms ω_η on M_η (via the diffeomorphisms) with symplectic forms on M_{η^0}, we have*

$$\omega_\eta = \omega_{\eta^0} + <\eta - \eta^0, c>$$

where $c \in \Omega^2(M_{\eta^0}) \otimes \mathbf{t}$ is a closed differential form. (Informally, the symplectic form on a reduced space M_η depends linearly on η.)

Corollary *The symplectic volume* $\mathrm{vol}(M_\eta)$ *of a family of symplectic quotients is a polynomial function of η in a sufficiently small neighbourhood of a regular value η^0.*

We shall prove this theorem starting from the following:

10.2   Normal Form Theorem

The proof presented here is adapted from [3], Proposition 40.1.

Proposition 10.4 (Normal form theorem) *We give a normal form for the T action, the symplectic structure and the moment map in a neighbourhood of $\Phi^{-1}(\eta^0)$ for any regular value η^0 of Φ.*

The role played by this result is analogous to the Darboux theorem.

Proof We begin with the following observation. If (M, ω) has a Hamiltonian T action and $H \le T$, then the moment polytope for Φ_H is obtained as follows. By reduction in stages, if $\zeta \in \mathbf{h}^*$ then $M_\zeta := \Phi_H^{-1}(\zeta)/H$ is a family of symplectic manifolds with Hamiltonian action of T/H. The moment polytopes are

$$\Phi_{T/H}(M_\zeta) = \{\xi \in \Phi_T(M) : \pi_H(\xi) = \zeta\}. \qquad \square$$

Example 10.5

$$M = \mathbb{C}P^2$$

The images $\Phi_{T/H}(M_\zeta)$ form a family of intervals of length $1 - \zeta$. Thus M_ζ is a 2-sphere with symplectic area $1 - \zeta$ (in other words, the symplectic form on M_ζ varies linearly with ζ, as stated in the Duistermaat–Heckman theorem).

Recall that if P is a manifold with a free action of G, then $P \to M = P/G$ inherits the structure of a principal G-bundle. A connection on G is a 1-form

$$\theta \in \Omega^1(P) \otimes \mathbf{t}$$

for which

- $(R_g)^* \theta = \mathrm{Ad}(g^{-1})\theta$
- $\theta(X^\#) = X$ for any $X \in \mathbf{g}$.

Example 10.6 If $U(1) \to P \to M$ is a principal $U(1)$-bundle, then a connection is a 1-form θ for which

- $(R_g)^* \theta = \theta$ (θ is invariant under the $U(1)$ action)
- $\theta(X^\#) = X$ for any $X \in i\mathbb{R} := \mathrm{Lie}(U(1))$.

Example 10.7 If $U(1)^n \to P \to M$ is a principal $U(1)^n$ bundle, then a connection is a collection of 1-forms $(\theta_1, \ldots, \theta_n)$ on P invariant under the action of $T = U(1)^n$ and for which $\theta_j(\xi_k^\#) = \delta_{jk}$ if $\xi_k^\#$ is the vector field generated by the kth copy of $U(1)$.

We then have the following.

Theorem 10.8 *Let (M, ω) be a symplectic manifold equipped with a Hamiltonian action of $T = U(1)^n$. Use $(\theta_1, \ldots, \theta_n)$ to define a connection on the bundle $\Phi^{-1}(\eta^0) \times \mathbb{R}^n$ over $\Phi^{-1}(\eta^0)/U(1)$ (where we have identified $\mathbf{t}$ with $\mathbb{R}^n$). Let ω_{η^0} be the symplectic form on $M_{\eta^0} = \Phi^{-1}(\eta^0)/T$ and define a symplectic structure on $\Phi^{-1}(\eta^0) \times \mathbb{R}^n$ by*

$$\omega = \pi^* \omega_{\eta^0} - d\left(\sum_{j=1}^{n} t_j \theta_j\right)$$

where t_j are coordinates on $\mathbb{R}^n \cong \mathbf{t}^*$ corresponding to the coordinates on $\mathbf{t}$ used to define the θ_j.

This theorem gives a description of the symplectic form and moment map and T action in a neighbourhood of η^0.

The action of T is defined by the action on $\Phi^{-1}(\eta^0)$. Then there is a symplectomorphism from a tubular neighbourhood of $\Phi^{-1}(\eta^0)$ in M to a tubular neighbourhood of $\Phi^{-1}(\eta^0) \times \{0\}$ in $\Phi^{-1}(\eta^0) \times \mathbb{R}^n$.

Lemma 10.9 *With the symplectic form ω on $\Phi^{-1}(\eta^0) \times \mathbb{R}^n$, the moment map is*

$$\Phi' : (p, (t_1, \ldots, t_n)) \mapsto -(t_1, \ldots, t_n).$$

Proof We have

$$\iota_{\xi_j} d\left(\sum_k t_k \theta_k\right) = -d\iota_{\xi_j}\left(\sum_k t_k \theta_k\right) = -dt_j$$

(since $L_{\xi_j} \theta_k = 0$). $\qquad \square$

Remark The isomorphism with a tubular neighbourhood is not canonical. It depends on the choice of a connection $(\theta_1, \ldots, \theta_n)$.

Remark If 0 is a regular value of the moment map, an analogous statement is true for the normal form for the action of a nonabelian group.

10.3   Duistermaat–Heckman Theorem, Version I

Theorem 10.10 (Duistermaat–Heckman) *If η^0 is a regular value of $\Phi : M \to \mathbf{t}^*$, then for η in a sufficiently small neighbourhood of η^0, $M_\eta \cong M_{\eta^0}$ and*

$$\omega_\eta = \omega_{\eta^0} + \sum_{j=1}^{n} (\eta - \eta^0)_j d\theta_j$$

Here we have decomposed $\eta = (\eta_1, \ldots, \eta_n) \in \mathbb{R}^n = \mathbf{t}^*$. In other words, the symplectic form varies linearly in the parameters η_j.

Proof If $\Phi^{-1}(\eta) \times \{\eta\}$ is in the open neighbourhood of $\Phi^{-1}(\eta^0) \times \{\eta^0\}$ which is identified diffeomorphically with a tubular neighbourhood of $\Phi^{-1}(\eta^0)$ in M, then

$$M_\eta = \Phi^{-1}(\eta)/T = (\Phi^{-1}(\eta^0) \times \{\eta\})/T$$

$$= \Phi^{-1}(\eta^0)/T \times \{\eta\} = M_{\eta^0}.$$

The symplectic form on M_η pulls back on

$$\Phi^{-1}(\eta^0) \times \{\eta\}$$

to the restriction

$$\omega_{\eta^0} - d\left(\sum_j (\eta_j - \eta_j^0)\theta_j \right).$$

But now η is a constant so the symplectic form pulls back to

$$\omega_{\eta^0} - \sum_j (\eta_j - \eta_j^0)d\theta_j.$$

Now $c_j := d\theta_j$ is a closed 2-form on $\Phi^{-1}(\eta^0) \times \mathbb{R}^n$. In fact the form c_j is pulled back from a 2-form on the symplectic quotient M_{η^0}. It is a representative in de Rham cohomology for the first Chern class of a line bundle L_j over M_{η^0}. If v_j is the element of $\mathbf{t}^*$ defining the coordinate t_j on $\mathbb{R}^n$, take $\rho_j = \exp(v_j) \in \Lambda^W = \mathrm{Hom}(T, U(1))$. The space

$$L_j = \Phi^{-1}(\eta^0) \times_{T,\rho_j} \mathbb{C}$$

is a line bundle over M_{η^0}. Then the 1-form θ_j is a connection on the line bundle L_j so $d\theta_j$ is its curvature. $\qquad\square$

Proposition 10.11 *The pushforward* $\Phi_*(\omega^N/N!)$ *at* $\eta \in \mathbf{t}^*$ *is equal to the symplectic volume of* M_η *multiplied by* $\mathrm{vol}(T)$.

Proof For a smooth function f on $\mathbf{t}$,

$$\int_{\eta \in \mathbf{t}^*} \Phi_*\left(\frac{\omega^N}{N!}\right) f(\eta) = \int_{m \in M} \frac{\omega^N}{N!} f(\Phi(m))$$

$$= \int_M (\exp\omega) f(\Phi(m)).$$

Choose f supported on the neighbourhood $U \in \mathbf{t}^*$. Then the integral becomes

$$\int_{(p,\eta) \in \Phi^{-1}(\eta^0) \times U} \exp\left(\pi^* \omega_{\eta^0} - d(\eta - \eta^0, \theta)\right) f(\eta)$$

$$= \int_{(p,\eta) \in \Phi^{-1}(\eta^0) \times \mathbf{t}^*} \exp(\pi^* \omega_{\eta^0}) \exp\left(-(d\eta, \theta) - (\eta - \eta^0, d\theta)\right) f(\eta).$$

The measure on $\mathbf{t}^*$ comes from $d\eta_1 \wedge \ldots \wedge d\eta_n$ in $d\eta_1 \wedge \ldots \wedge d\eta_n \wedge \theta_1 \wedge \ldots \wedge \theta_n$.

We get this by expanding $\exp\{-(d\eta, \theta)\}$. We evaluate the integral over $\Phi^{-1}(\eta^0) \times \{\eta\}$, to get

$$\int \theta_1 \wedge \ldots \wedge \theta_n \exp(\pi^* \omega_{\eta^0}) \exp\left(-(\eta - \eta^0), d\theta\right)$$

$$= \mathrm{vol}(M_\eta)\mathrm{vol}(T).$$

(since $\mathrm{vol}(T) = \int_T \theta_1 \wedge \ldots \wedge \theta_n$.) The remaining integral is over $\eta \in \mathbf{t}^*$, so it is $\int_{\eta \in \mathbf{t}^*} \mathrm{vol}(M_\eta)\mathrm{vol}(T)g(\eta)$ so

$$\mathrm{vol}_\omega(M) = \int_{\mathbf{t}^*} \Phi_*(\omega^N/N!) = \mathrm{vol}(T) \int_{\eta \in \mathbf{t}^*} \mathrm{vol}_\omega(M_\eta).$$

This follows by applying the definition of pushforward to $g : \mathbf{t}^* \to \mathbb{R}$ given by $g(x) = 1$. $\qquad\square$

Corollary 10.12 *If* M^{2N} *is a toric manifold (acted on effectively by* $U(1)^N$*), then its symplectic volume is equal to the Euclidean volume of its Newton polytope (the image of the moment map for the torus action).*

The proof uses the fact that for a toric manifold, the symplectic quotient at η is a point if η is in the image of the moment map, and it is empty otherwise. Notice that the pushforward $\Phi_*(\frac{\omega^N}{N!})$ of the Liouville measure is supported on the (compact) polytope $\Phi(M)$, but it encodes more information about M than just the polytope.

Proposition 10.13 *The n-form $\Phi_*(\frac{\omega^N}{N!})$ is a polynomial of degree $\leq N$ on sufficiently small neighbourhoods of regular values of Φ.*

Proof This result follows immediately from Theorem 10.10. We learn from this theorem that $\Phi_*(\frac{\omega^N}{N!})$ is polynomial on any connected component of the set of regular values of Φ, and it and its derivatives may have discontinuities on the hyperplanes (walls) consisting of critical values of Φ.   □

Example 10.14 Consider the adjoint action of the maximal torus T on the adjoint orbit of an element in $su(3)$. (See, for example, the book [2] by Guillemin, Lerman and Sternberg.) The measure $\Phi_*(\frac{\omega^N}{N!})$ is Euclidean measure multiplied by a piecewise linear function characterized by

1. $\Phi_*(\frac{\omega^N}{N!}) = 0$ on the boundary of $\Phi(O_\lambda)$
2. On the region adjacent to the boundary, $\Phi_*(\frac{\omega^N}{N!})$ is proportional to the Euclidean distance to the boundary.
3. $\Phi_*(\frac{\omega^N}{N!})$ is constant on the interior triangle (the component of the complement of the walls containing the centre of the hexagon).

Remarks on torus actions

Proposition 10.15 *The orbits of a Hamiltonian torus action are isotropic.*

Proof $\omega(X_m^\#, Y_m^\#) = 0$ for any $X, Y \in \mathbf{t}$.   □

Corollary *If M^{2N} is a toric manifold (acted on by $T \cong U(1)^N$), then*

$$M \xrightarrow{\Phi} B$$

has the property that $\Phi^{-1}(b)$ is a Lagrangian submanifold for any regular value b of Φ. In other words, the map $\Phi^{-1}\big(\mathrm{Int}(B)\big) \to B$ is a fibration with Lagrangian fibres isomorphic to $U(1)^N$.

This is a special case of the Liouville–Arnol'd theorem. See, for example, the book [4] by Arnol'd.

10.4   Computation of Pushforward of Liouville Measure on a Symplectic Vector Space

Recall that $V \cong \mathbb{C}^N$ is a symplectic vector space acted on linearly by a torus T; in other words,

$$V \cong \oplus_\beta \mathbb{C}_\beta$$

where $\mathbb{C}_\beta$ is acted on by

$$\rho_\beta = \exp(2\pi i \beta) \in \mathrm{Hom}(T, U(1))$$

for

$$\beta \in \Lambda^W = \mathrm{Hom}(\Lambda^I, \mathbb{Z}) \subset \mathfrak{t}^*.$$

Here Λ^I denotes the integer lattice (the kernel of the exponential map), and Λ^W denotes the weight lattice (defined here). We saw that the moment map was

$$\Phi(z_1, \ldots, z_N) = -\frac{1}{2}\sum_j |z_j|^2 \beta_j$$

$$\omega = \frac{i}{2}\sum_j dz_j \wedge d\bar{z}^j = \sum_j dx_j \wedge dy_j.$$

Denote the moment map for the linear action by

Lemma 10.16 *The pushforward of Liouville measure under the moment map Φ is*

$$\Phi_*(\frac{\omega^N}{N!})(\xi) = H_{\bar{\beta}}(\xi)dt_1 \wedge \cdots \wedge dt_\ell$$

where this function is defined by

$$H_{\bar{\beta}}(\xi) = \mathrm{vol}\{(s_1, \ldots, s_N) \in (\mathbb{R}^+)^N : \xi = -\sum_{j=1}^{N} s_j \beta_j\}.$$

Here, the number of equations is ℓ (the dimension of T) and the number of unknowns is N. The function $H_{\bar{\beta}}$ is piecewise polynomial of degree $N - \ell$. The pushforward of Lebesgue measure on $\mathbb{R}^N$ under the moment map Φ is the function $H_{\bar{\beta}}(\xi)$ multiplied by Euclidean measure on $\mathbb{R}^\ell$.

Let $\xi \in \mathfrak{t}^*$ and suppose that $L : \mathbb{C}^N \to \mathfrak{t}^*$ is the map

$$L(s_1, \ldots, s_N) = \sum_j s_j \beta_j.$$

For toric manifolds ($N = \ell$), $H_{\bar{\beta}}$ is the characteristic function of the image of the moment map of the torus action (also called the Newton polytope).

In the case $\mathbb{C} \to \mathbb{R}, z \mapsto \frac{1}{2}|z|^2$, on $\Psi_*(dx \wedge dy) = 2\pi ds$ when s is the coordinate on $\mathbb{R}$. This is because $dxdy = rdrd\theta = \frac{1}{2}d(r^2)d\theta$. So

$$\int f(\frac{1}{2}r^2)\frac{1}{2}d(r^2)d\theta = 2\pi \int f(r^2)d(\frac{1}{2}r^2)$$

so

$$\Psi_*(dxdy) = 2\pi ds.$$

So

$$\Phi = L \circ (\Psi_1, \ldots, \Psi_N)$$

where

$$\Psi_j : \mathbb{C} \to \mathbb{R}^+$$

is

$$\Psi_j(z) = \frac{1}{2}|z|^2.$$

So

$$\Phi_*(\omega^N/N!) = L_*((2\pi)^N ds_1 \wedge \ldots ds_N).$$

It is easy to check that

$$L_*(ds_1 \wedge \ldots ds_N)(\xi) =$$

$$\mathrm{vol}\left((s_1, \ldots, s_N) \in (\mathbb{R}^+)^N : \xi = -\sum_{j=1}^N s_j \beta_j\right).$$

Let us define a differential operator on $\mathbf{t}^*$ by

$$D_{\beta_j} = \beta_j(\frac{\partial}{\partial \xi_1}, \cdots, \frac{\partial}{\partial \xi_N}).$$

Then

$$\prod_{j=1}^N D_{\beta_j} H_{\bar{\beta}} = \delta(\xi)$$

so $H_{\bar{\beta}}$ is the fundamental solution of a differential equation with support on the cone $-C_{\bar{\beta}}$. A good reference for this material is the book [2] by Guillemin, Lerman and Sternberg.

Let M be a symplectic manifold equipped with the Hamiltonian action of a group G. As described in the previous chapter, the equivariant 2-form $\bar{\omega} \in \Omega_G^2(M)$ defined by

$$\bar{\omega}(X) = \omega + (\mu, X)$$

satisfies $D\bar{\omega} = 0$ and thus defines an element $[\bar{\omega}] \in H_G^2(M)$.

Theorem 10.17 (Duistermaat–Heckman theorem, version II) *Suppose M is a symplectic manifold of dimension $2n$ equipped with the Hamiltonian action of a torus T. Then for generic $X \in \mathbf{t}$, in the notation of Theorem 9.50, we have*

$$\int_M e^{i\bar{\omega}} = \int_M \frac{(i\omega)^n}{n!} e^{i\mu(m)(X)} = \sum_{F \in \mathcal{F}} e^{i\mu(F)(X)} \int_F \frac{e^{i\omega}}{e_F(X)}.$$

Proof Apply the abelian localization theorem (Theorem 9.50) to the class

$$\exp(i\bar{\omega}) \in H_G^*(M).$$

(For each component F of the fixed point set of T, the value of $\mu(F)$ is a constant.)

□

10.5 Stationary Phase Approximation

An alternative approach to this version of the Duistermaat–Heckman theorem ("exactness of the stationary phase approximation") is sketched as follows. Assume for simplicity that $T = U(1)$ and that the components F of the fixed point set are isolated points. By the equivariant version of the Darboux–Weinstein theorem [5], we may assume the existence of Darboux coordinates $(x_1, y_1, \ldots, x_n, y_n)$ on a coordinate patch U_F about F, for which

$$\mu(x_1, y_1, \ldots, x_n, y_n) = \mu(F) - \sum_j \frac{m_j}{2}(x_j^2 + y_j^2).$$

Thus the oscillatory integral over U_F tends (if we may replace U_F by $\mathbb{R}^{2n}$) to

$$\int_{m \in M} e^{i\omega} e^{i\mu(m)X} = \int_{\mathbb{R}^{2n}} i^n dx_1 dy_1 \ldots dx_n dy_n e^{i\mu(F)X} e^{-i\sum_j m_j(x_j^2 + y_j^2)/2}. \qquad (10.1)$$

Here X is a real parameter. The integral over $\mathbb{R}^{2n}$ is given by a standard Gaussian integral:

$$\int_{\mathbb{R}^{2n}} e^{i\omega} e^{i\mu X} = \frac{(2\pi)^n e^{i\mu(F)X}}{(\prod_j m_j) X^n} := S_F(X)$$

The *lemma of stationary phase* ([3], Sect. 33) asserts that the oscillatory integral $\int_{m \in M} e^{i\omega} e^{i\mu(m)X}$ over M has an asymptotic expansion as $X \to \infty$ given by

$$\int_{m \in M} e^{i\omega} e^{i\mu(m)X} = \sum_{F \in \mathcal{F}} S_F(X)(1 + O(1/X)) + O(X^{-\infty}).$$

The first version of the Duistermaat–Heckman theorem (Theorem 10.17) may thus be reformulated as the assertion that *the stationary phase approximation is exact* (in other words, the leading order term in the asymptotic expansion gives the exact answer for any value of the parameter X).

10.6   The Kirwan Map

Suppose M is a compact symplectic manifold equipped with a Hamiltonian action of a compact Lie group G. Suppose 0 is a regular value of the moment map μ. There is a natural map $\kappa : H_G^*(M) \to H^*(M_{\text{red}})$ defined by

$$\kappa : H_G^*(M) \mapsto H_G^*(Z_0) \cong H^*(M_{\text{red}})$$

where $Z_0 := \mu^{-1}(0)$. This map is obviously a ring homomorphism.

Theorem 10.18 (Kirwan) *The map κ is surjective.*

The proof of this theorem ([6], Sects. 5.4 and 8.10; see also Sect. 6 of [7]) uses the Morse theory of the "Yang–Mills function" $|\mu|^2 : M \to \mathbb{R}$ to define an equivariant stratification of M by strata S_β which flow under the gradient flow of $-|\mu|^2$ to a critical set C_β of $|\mu|^2$. One shows that the function $|\mu|^2$ is *equivariantly perfect*, in other words, that the Thom–Gysin (long) exact sequence in equivariant cohomology decomposes into short exact sequences, so that one may build up the cohomology as

$$H_G^*(M) \cong H_G^*(\mu^{-1}(0)) \oplus \bigoplus_{\beta \neq 0} H_G^*(S_\beta).$$

Here, the stratification by S_β has a partial order $>$; thus one may define an open dense set $U_\beta = M - \cup_{\gamma > \beta} S_\gamma$ of all points that flow into S_β. This includes the open dense stratum S_0 of points that flow into $\mu^{-1}(0)$. Note that the stratum S_0 retracts onto $\mu^{-1}(0)$). The equivariant Thom–Gysin sequence is

$$\cdots \to H_G^{n-2d(\beta)}(S_\beta) \xrightarrow{i_{\beta*}} H_G^n(U_\beta) \to H_G^n(U_\beta \setminus S_\beta) \to \cdots .$$

To show that the Thom–Gysin sequence splits into short exact sequences, it suffices to know that the maps $(i_\beta)_*$ are injective. The map $i_\beta^*(i_\beta)_*$ is multiplication by the equivariant Euler class e_β of the normal bundle to S_β. (See, for example, Chap. 9.) Injectivity follows because this equivariant Euler class is not a zero divisor (see [6], Theorem 5.4 for the proof).

Atiyah and Bott [8] use a similar argument in an infinite-dimensional context to define a stratification of the infinite-dimensional space of all connections on a compact orientable 2-manifold Σ, using the Yang–Mills functional $\int_\sigma |F_A|^2$ (which is equivariant with respect to the action of the gauge group). This stratification is used to compute the dimensions of the cohomology groups of the spaces of gauge equivalence classes of flat connections.

10.7 Nonabelian Localization

Witten in [9] gave a result (the *nonabelian localization principle*) that related inter-section pairings on the symplectic quotient M_{red} of a (compact) manifold M to data on M itself. Since $\kappa : H_G^*(M) \to H^*(M_{\text{red}})$ is a surjective ring homomorphism, all intersection pairings are given in the form $\int_{M_{\text{red}}} \kappa(\eta)$ for some $\eta \in H_G^*(M)$.

In the paper [9], Witten regards the equivariant cohomology parameter $X \subset \mathfrak{g}$ as an integration variable and seeks to compute the asymptotics in $\epsilon > 0$ of

$$\int_{X \in \mathfrak{g}} dX e^{-\epsilon |X|^2/2} \int_M \eta(X) e^{i\omega} e^{i(\mu, X)}. \tag{10.2}$$

He finds that the expression (10.2) has an asymptotic expansion as $\epsilon \to 0$ of the form

$$\int_{M_{\text{red}}} e^{\epsilon \Theta} e^{i\omega_{\text{red}}} \kappa(\eta) + O\left(p(\epsilon^{-1/2}) e^{-\frac{b}{2\epsilon}}\right) \tag{10.3}$$

where b is the smallest nonzero critical value of $|\mu|^2$, p is a polynomial, and Θ is a particular element of $H^4(M_{\text{red}})$ (the image $\kappa(\beta)$ of the element $\beta \in H_G^*(M)$ specified by $\beta : X \in \mathfrak{g} \mapsto -|X|^2/2$). Recall κ is the Kirwan map.

10.8 The Residue Formula

A related result is the *residue formula*, Theorem 8.1 of [10].

We define the residue on meromorphic functions of the form $\frac{e^{i\lambda X}}{X^N}$ when $\lambda \neq 0$ (for $0 < N \in \mathbb{Z}$) by

$$\text{Res}(\frac{e^{i\lambda X}}{X^N}) = \text{Res}_{X=0} \frac{e^{i\lambda X}}{X^N}, \lambda > 0;$$

$$= 0, \qquad \lambda < 0.$$

More generally, the residue is specified by certain axioms (see [10], Proposition 8.11), and may be defined as a sum of iterated multivariable residues $\text{Res}_{X_1=\lambda_1} \ldots \text{Res}_{X_l=\lambda_l}$ for a suitably chosen basis of $\mathbf{t}$ yielding coordinates $X_1, \ldots, X_l$ (see Proposition 3.2 in [11]).

Theorem 10.19 ([10], corrected as in [12])
Let $\eta \in H_G^*(M)$ induce $\eta^0 \in H^*(M_{\text{red}})$. Then we have

$$\int_{M_{\text{red}}} \kappa(\eta) e^{i\omega_{\text{red}}} = n_0 C^G \text{Res}\left(\mathcal{D}^2(X) \sum_{F \in \mathcal{F}} H_F^\eta(X)[dX]\right), \tag{10.4}$$

where n_0 is the order of the stabilizer in G of a generic element of $\mu^{-1}(0)$, and the constant C^G is defined by

$$C^G = \frac{(-1)^{s+n_+}}{|W|\mathrm{vol}(T)}. \tag{10.5}$$

We have introduced $s = \dim G$ and $l = \dim T$; here $n_+ = (s-l)/2$ is the number of positive roots.[1] Also, $\mathcal{F}$ denotes the set of components of the fixed point set of T, and if F is one of these components then the meromorphic function H_f^η on $\mathbf{t} \otimes \mathbb{C}$ is defined by

$$H_f^\eta(X) = e^{i\mu(F)(X)} \int_F \frac{i_F^* \eta(X) e^{i\omega}}{e_F(X)} \tag{10.6}$$

and the polynomial $\mathcal{D} : \mathbf{t} \to \mathbb{R}$ is defined by $\mathcal{D}(X) = \prod_{\gamma>0} \gamma(X)$, where γ runs over the positive roots of G.

The main ingredients in the proof of Theorem 10.19 are the normal form theorem (see Sect. 10.2) and the abelian localization theorem (Theorem 9.50). We outline a proof as follows. First (following Martin [13]), we may reduce to symplectic quotients by the action of the maximal torus T:

Proposition 10.20 ([13]) *We have*

$$\int_{\mu^{-1}(0)/G} \kappa(\eta e^{i\bar\omega}) = \frac{1}{|W|} \int_{\mu^{-1}(0)/T} \kappa(\mathcal{D}\eta e^{i\bar\omega}) = \frac{(-1)^{n_+}}{|W|} \int_{\mu_T^{-1}(0)/T} \kappa(\mathcal{D}^2 \eta e^{i\bar\omega}).$$

Proof We need to prove the result only for torus actions. A sketch of the proof when $G = U(1)$ [14] follows: We write

$$\eta = D\left(\frac{\theta\eta}{d\theta - X}\right).$$

Suppose 0 is a regular value of μ. Then $\mu^{-1}(\mathbb{R}^+)$ is a manifold with boundary $\mu^{-1}(0) := Z_0$. One may show [10] that

$$\mathrm{Res}_{X=0} \int_{Z_0} \frac{\theta\eta e^{i\bar\omega}}{X - d\theta} = \int_{Z_0/G} \kappa(\eta e^{i\bar\omega}). \tag{10.7}$$

In the $U(1)$ case the map κ may be written as

$$\kappa : \eta \mapsto \mathrm{Res}_{X=0} p_* \frac{\theta\eta}{X - d\theta}, \tag{10.8}$$

---

[1] Here, the roots of G are the nonzero weights of its complexified adjoint action. We fix the convention that weights $\beta \in \mathbf{t}^*$ satisfy $\beta \in \mathrm{Hom}(\Lambda^I, \mathbb{Z})$ rather than $\beta \in \mathrm{Hom}(\Lambda^I, 2\pi\mathbb{Z})$ (where $\Lambda^I = \mathrm{Ker}(\exp : \mathbf{t} \to T)$ is the integer lattice). This definition of roots differs by a factor of 2π from the definition used in [10].

(where $p : Z_0 \rightarrow Z_0/G$ is projection, so that the pushforward p_* is integration over the fibre of p). Applying the equivariant Stokes' theorem to $\mu^{-1}(\mathbb{R}^+)$ and then taking the residue at $X = 0$, we find that

$$\text{Res}_{X=0} \int_{Z_0} \frac{\theta\eta}{d\theta - X} - \sum_{F \in \mathcal{F} : \mu(F) > 0} \text{Res}_{X=0} e^{i\mu(F)X} \int_F \frac{\eta e^{i\omega}}{e_F(X)} = 0, \qquad (10.9)$$

which is exactly the $U(1)$ case of the residue formula.

Nonabelian localization has had two major applications thus far. The first is that the residue formula has been used in [12] to give a proof of formulas for intersection numbers on moduli spaces of vector bundles on Riemann surfaces. Some of the background underlying these results is described in Chap. 12. The second is that nonabelian localization underlies some proofs (see, e.g. [11, 15]) of a conjecture of Guillemin and Sternberg [16] that "quantization commutes with reduction": in other words, that the G-invariant part of the quantization of a symplectic manifold equipped with a Hamiltonian G action is isomorphic to the quantization of the reduced space M_{red}. See Chap. 11 of this volume. For an expository account and references on results about this conjecture of Guillemin and Sternberg, see the survey article by Sjamaar [17].

10.9 The Residue Formula by Induction

Guillemin and Kalkman [18] and independently Martin [13] have given an alternative version of the residue formula which uses the one-variable proof inductively.

Theorem 10.21 (Guillemin–Kalkman; Martin) *Suppose M is a symplectic manifold acted on by a torus T in a Hamiltonian fashion, and $\eta \in H_T^*(M)$. Then*

$$\int_{M_{\text{red}}} \kappa(\eta) = \sum_i {}' \int_{(M_i)_{\text{red}}} \kappa_i (\text{Res}_i \eta).$$

Here, M_i is the fixed point set of a one-parameter subgroup T_i of T (so that $\mu_T(M_i)$ are critical values of μ_T): it is a symplectic manifold equipped with a Hamiltonian action of T/T_i and with the natural map $\kappa_i : H_{T/T_i}^*(M_i) \rightarrow H^*((M_i)_{\text{red}})$ (the Kirwan map for the action of the group T/T_i). The map $\text{Res}_i : H_T^*(M) \rightarrow H_{T/T_i}^*(M_i)$ is defined by

$$\text{Res}_i \eta = \text{Res}_{X_i=0}(i_{M_i}^* \eta) \qquad (10.10)$$

where i_{M_i} is the inclusion map,

$$i_{M_i}^* \eta \in H_T^*(M_i) = H_{T/T_i}^*(M_i) \otimes H_{T_i}^*$$

and $X_i \in \mathfrak{t}_i^*$ is a basis element for $\mathfrak{t}_i^*$.

The sum in Theorem 10.21 is over those T_i and M_i for which a (generic) ray in $\mathbf{t}^*$ from 0 to the complement of $\mu_T(M)$ intersects $\mu_T(M_i)$. Different components M_i and groups T_i will contribute depending on the choice of the ray.

References

1. J.J. Duistermaat, G. Heckman, On the variation in the cohomology of the symplectic form of the reduced phase space. Invent. Math. **69**, 259–268 (1982). Addendum, **72**, 153–158 (1983)
2. V. Guillemin, E. Lerman, S. Sternberg, *Symplectic Fibrations and Multiplicity Diagrams* (Cambridge University Press, 1996)
3. V. Guillemin, S. Sternberg, *Symplectic Techniques in Physics* (Cambridge University Press, 1990)
4. V.I. Arnold, *Mathematical Methods of Classical Mechanics*. Graduate Texts in Mathematics (Springer, 1978)
5. A. Weinstein, Symplectic manifolds and their Lagrangian submanifolds. Adv. Math. **6**, 329–346 (1970)
6. F. Kirwan, *Cohomology of Quotients in Symplectic and Algebraic Geometry* (Princeton University Press, 1984)
7. F. Kirwan, The cohomology rings of moduli spaces of vector bundles over Riemann surfaces. J. Amer. Math. Soc. **5**, 853–906 (1992)
8. M.F. Atiyah, R. Bott, The Yang–Mills equations over Riemann surfaces. Philos. Trans. Roy. Soc. Lond. A **308**, 523–615 (1982)
9. E. Witten, Two dimensional gauge theories revisited. J. Geom. Phys. **9**, 303–368 (1992)
10. L.C. Jeffrey, F.C. Kirwan, Localization for nonabelian group actions. Topology **34**, 291–327 (1995)
11. L.C. Jeffrey, F. Kirwan, Localization and the quantization conjecture. Topology **36**, 647–693 (1997)
12. L.C. Jeffrey, F.C. Kirwan, Intersection theory on moduli spaces of holomorphic bundles of arbitrary rank on a Riemann surface. Ann. Math. **148**, 109–196 (1998)
13. S.K. Martin, Symplectic quotients by a nonabelian group and by its maximal torus, arXiv:math/0001002 (2000)
14. J. Kalkman, Cohomology rings of symplectic quotients. J. Reine Angew. Math. **458**, 37–52 (1995)
15. M. Vergne, Multiplicity formulas for geometric quantization I. Duke Math. J. **82**, 143–179 (1996)
16. V. Guillemin, S. Sternberg, Geometric quantization and multiplicities of group representations. Invent. Math. **67**, 515–538 (1982)
17. R. Sjamaar, Symplectic reduction and Riemann–Roch formulas for multiplicities. Bull. Amer. Math. Soc. **33**, 327–338 (1996)
18. V. Guillemin, J. Kalkman, The Jeffrey–Kirwan localization theorem and residue operations in equivariant cohomology. J. Reine Angew. Math. **470**, 123–142 (1996)

Chapter 11
Geometric Quantization

This chapter describes geometric quantization. The motivation for this mathematical is to mimic quantum mechanics, where a manifold (the "classical phase space", parametrizing position and momentum) is replaced by a vector space with an inner product; in other words, a Hilbert space (the "space of wave functions"). Functions on the manifold ("observables") are replaced by endomorphisms of the vector space.

In short, geometric quantization replaces a symplectic manifold (the classical phase space) by a vector space with inner product (the physical Hilbert space).

If the symplectic manifold has dimension n, the quantization $\mathcal{H}$ should consist of "functions of half the variables." A prototype is $M = \mathbb{R}^2$ with coordinates q (position) and p (momentum).

One way to do this is to let $\mathcal{H}$ be holomorphic functions in $p + iq$ (this is called a complex polarization). Alternatively, [1] we could use a real polarization (a map π to a manifold of half the dimension, with fibres of Lagrangian submanifolds) and define $\mathcal{H}$ to be functions' covariant constant along the fibres of the polarization (in the $\mathbb{R}^2$ example, an example would be functions of p or of q). Geometric quantization with a real polarization gives a basis consisting of sections of the prequantum line bundle covariant constant along the fibres.

The layout of the chapter is as follows. In Sect. 11.1, we describe holomorphic line bundles over a complex manifold. In Sect. 11.2, we describe the geometric quantization of $\mathbb{C}P^1$. In Sect. 11.3, we give a link to representation theorem. In Sect. 11.4, we describe the Bott–Borel–Weil theorem. Finally in Sect. 11.5, we outline the representation theory of $SU(2)$.

© The Author(s), under exclusive licence to Springer Nature Switzerland AG 2019
S. Dwivedi et al., *Hamiltonian Group Actions and Equivariant Cohomology*,
SpringerBriefs in Mathematics,
https://doi.org/10.1007/978-3-030-27227-2_11

11.1   Holomorphic Line Bundle over a Complex Manifold

Definition 11.1 A complex line bundle over a smooth manifold M is specified by
an open cover $\{U_\alpha\}$ on M and transition functions $g_{\alpha\beta} : U_\alpha \cap U_\beta \to \mathbb{C}^*$. The line
bundle is defined as

$$L = \cup_\alpha U_\alpha \times \mathbb{C}/\sim$$

where $(x, z_\alpha) \sim (x, z_\beta)$ if $z_\alpha = g_{\alpha\beta}(x)z_\beta$.

Definition 11.2 Suppose M is a complex manifold. The line bundle is holomorphic
if the transition functions $g_{\alpha\beta}$ are holomorphic.

Suppose the complex structure is compatible with the symplectic structure:

$$\omega(JX, JY) = \omega(X, Y)$$

for $X, Y \in T_x M$ and J the corresponding almost complex structure. We assume that
the almost complex structure J is integrable (in other words, it comes from a complex
manifold, as in Remark 1.15 of Chap. 1. In this situation, the manifold is called a
Kähler manifold.

Definition 11.3 A section s of L is a collection of maps $s_\alpha : U_\alpha \to \mathbb{C}$ satisfying
$s_\alpha(z) = g_{\alpha\beta}(z)s_\beta(z)$ for $z \in U_\alpha \cap U_\beta$. This is well defined, since $\frac{\partial}{\partial \bar{z}_j} g_{\alpha\beta} = 0$ so on
$U_\alpha \cap U_\beta$, $\frac{\partial}{\partial \bar{z}_j} s_\alpha = 0$ if and only if $\frac{\partial}{\partial \bar{z}_j} s_\beta = 0$. We will denote the sections of L by
$\Gamma(L)$.

Definition 11.4 The complex tangent space and cotangent space are defined as follows:

$$T_\mathbb{C}M = TM \otimes \mathbb{C}$$

$$T_\mathbb{C}^*M = T^*M \otimes \mathbb{C}$$

In local complex coordinates z_j, a basis for $T_\mathbb{C}^*M$ is $\{dz_j, d\bar{z}_j\}$, $j = 1, \ldots, n$.

Definition 11.5 The holomorphic and antiholomorphic cotangent spaces of a complex manifold M are denoted by

$$T_\mathbb{C}^*M = (T^*)^{(1,0)}M \oplus (T^*)^{(0,1)}M$$

$$= (T^*)'M \oplus (T^*)''M$$

where in local complex coordinates $(T^*)''M$ is spanned by $\{d\bar{z}_j\}$ and $(T^*)'M$ is
spanned by $\{dz_j\}$.

Definition 11.6 ($\bar{\partial}$-*operator on functions on* M) Choose local complex coordinates
$z_1, \ldots, z_n$ on the U_α and define

$$\bar{\partial} : C^{\infty}(U_{\alpha}) \to \Omega^{0,1}(U_{\alpha})$$

$$\bar{\partial} f = \sum_{j=1}^{n} \frac{\partial f}{\partial \bar{z}_j} d\bar{z}_j$$

Remark 11.7 We note that the definition of $\bar{\partial}$ is independent of the choice of complex coordinates—on the other hand, the quantization depends on the choice of complex structure or almost complex structure. If M has an almost complex structure that is not integrable, the definition of the quantization can be generalized using Dolbeault operators. See for example [2].

Definition 11.8 ($\bar{\partial}$ *operator on sections of L on M*) Suppose L is a holomorphic line bundle over M. Given a section $s : M \to L, s = \{s_{\alpha}\}$, define

$$\bar{\partial}s \in \Gamma((T^*)^{''}M \otimes L)$$

by

$$\bar{\partial}s = \bar{\partial}s_{\alpha}$$

on U_{α}.

This is well defined since $\bar{\partial}g_{\alpha\beta} = 0$.

Proposition 11.9 *Specifying a structure of holomorphic line bundle on a complex line bundle L is equivalent to specifying operators $\bar{\partial} : \Gamma(L) \to \Omega^{0,1}(M, L)$ and $\bar{\partial} : \Omega^{0,1}(M, L) \to \Omega^{0,2}(M, L)$ satisfying $\bar{\partial} \circ \bar{\partial} = 0$.*

Proof We have seen that a holomorphic line bundle determines a $\bar{\partial}$ operator. Conversely, given a complex line bundle L with $\bar{\partial}$, we can choose an open cover $\{U_{\alpha}\}$ with locally defined solutions $s_{\alpha} \in \Gamma(L|_{U_{\alpha}})$ to $\bar{\partial}s_{\alpha} = 0$, and $s_{\alpha}(x) \neq 0$ for all $x \in U_{\alpha}$. Define transition functions $g_{\alpha\beta} : U_{\alpha} \cap U_{\beta} \to \mathbb{C}^*$ by

$$g_{\alpha\beta} = s_{\alpha}s_{\beta}^{-1}.$$

Hence, $\bar{\partial}g_{\alpha\beta} = 0$ so $g_{\alpha\beta}$ gives L the structure of a holomorphic line bundle. $\qquad\square$

Definition 11.10 A prequantum line bundle with connection is a complex line bundle $L \to M$ equipped with a connection θ whose curvature is equal to the symplectic form.

Proposition 11.11 *Let (L, ∇) be a prequantum line bundle over M. Suppose M is equipped with a complex structure J compatible with ω (in other words, on M there are locally defined complex coordinates $\{z_j\}$). Then $\nabla : \Gamma(L) \to \Gamma(T^*M \otimes L)$ decomposes as*

$$\nabla = \nabla^{'} \oplus \nabla^{''}$$

where

$$\nabla' : \Gamma(L) \to \Gamma((T^*)'M \otimes L)$$

and

$$\nabla'' : \Gamma(L) \to \Gamma((T^*)''M \otimes L).$$

Note that ∇' and ∇'' depend on the almost complex structure J on M.

Proposition 11.12 *We may define a structure of holomorphic line bundle on L by defining ∇'' as a $\bar{\partial}$ operator: a section s of L is defined to be holomorphic if*

$$\nabla'' s = 0.$$

Definition 11.13 The quantization of the symplectic manifold (M, ω) equipped with the prequantum line bundle L with connection ∇ and complex structure J is

$$\mathcal{H} = H^0(M, L),$$

in other words the global holomorphic sections of L.

As noted in Remark 11.7, the quantization can be defined even if M is equipped with a complex structure which is not integrable.

Remark 11.14 If M is compact, the vector space $\mathcal{H}$ is a finite-dimensional complex vector space.

11.2 Quantization of $\mathbb{C}P^1 \cong S^2$

The two-sphere is

$$\mathbb{C}P^1 = \left\{ (z_0, z_1) \in \mathbb{C}^2 \setminus \{(0, 0)\} \right\} / \sim$$

$$= \{[z_0 : z_1]\}$$

(using the standard notation for projective spaces, where $(z_0, z_1) \sim (\lambda z_0, \lambda z_1)$ for $\lambda \in \mathbb{C}$ with $\lambda \neq 0$). The hyperplane line bundle over $\mathbb{C}P^1$ is defined in [3], Chapter 0. The fibre above $[z_0 : z_1]$ in this bundle is $L_{[z_0:z_1]} = \left\{ f : \{\lambda(z_0, z_1)\} \to \mathbb{C} \right\}$, (in other words, linear functions on the line $\{(\lambda z_0, \lambda z_1)\}$ where $\lambda \in \mathbb{C}$ and z_0 and z_1 are constant). Such f satisfy

$$f(\lambda z_0, \lambda z_1) = \lambda f(z_0, z_1)$$

The dual of the hyperplane line bundle is the tautological line bundle, for which the fibre above $[z_0 : z_1]$ is

$$L^*_{[z_0:z_1]} = \Big\{ \{(\lambda z_0, \lambda z_1)\} : \lambda \in \mathbb{C} \Big\}.$$

This is the collection of complex lines through the point $(z_0, z_1) \in \mathbb{C}^2$. The kth power of the hyperplane line bundle is

$$L^k_{[z_0:z_1]} = \Big\{ f : \{(\lambda z_0, \lambda z_1)\} \to \mathbb{C} : f(\lambda z_0, \lambda z_1) = \lambda^k f(z_0, z_1) \Big\}$$

in other words, f is a polynomial of degree k on the line through $(z_0, z_1) \in \mathbb{C}^2 \setminus \{0\}$. Its zeroth power is the trivial bundle $L^0 = \mathbb{C}P^1 \times \mathbb{C}$.

11.2.1 Global Holomorphic Sections

The space $H^0(L)$ is spanned by the restrictions to $\mathbb{C}^2 \setminus \{0\}$ of the linear functions on $\mathbb{C}^2$. This is a complex vector space of dimension 2. The space $H^0(L^k)$ is spanned by the restrictions to $\mathbb{C}^2 \setminus \{0\}$ of the polynomials of degree k on $\mathbb{C}^2$:

$$f(z_0, z_1) = \sum_{j=0}^{k} a_j z_0^j z_1^{k-j}. \tag{11.1}$$

This is a complex vector space of dimension $k + 1$.

11.3 Link to Representation Theory

Suppose a (compact) group G acts on M (from the left), preserving the complex structure J as well as the symplectic structure (in other words, for each $g \in G$, $L_g : M \to M$ is a holomorphic diffeomorphism).

Suppose the G action lifts to an action on the total space L of a prequantum line bundle which preserves the connection ∇, and that this action is linear in the fibres: in other words,

$$L_g : \pi^{-1}(m) \to \pi^{-1}(gm) \tag{11.2}$$

is a linear map.

Proposition 11.15 *In this situation, the G action on M defines an action of G on $H^0(L)$ (from the right).*

Define $(s \cdot g)(m) = s(g(m))$, in other words $s \cdot g = s \circ L_g$. Thus, since L_g is a holomorphic diffeomorphism, the composition $s \circ L_g$ is a holomorphic section.

Proposition 11.16 *The action of G on the space of holomorphic sections is linear. Thus $\mathcal{H}$ is a linear representation of G.*

Proof If s_1 and s_2 are holomorphic sections of L, then $(s_1 + s_2) \cdot g = s_1 \cdot g + s_2 \cdot g$. This follows from the linearity of the action of g on fibres of the prequantum line bundle.   □

Remark 11.17 We shall often use a conjugation-invariant inner product on **g** (such as the Killing form) and its restriction to **t** to identify **t** with **t***. Thus, a weight $\lambda \in \Lambda^W$ will sometimes be viewed as an element of **t**, although strictly speaking $\Lambda^W \subset \mathbf{t}^*$.

Proposition 11.18 *Let M be a symplectic manifold acted on by T, and suppose ω is an integral symplectic form. Then the weights $\beta \in \mathbf{t}^*$ of the representation of T on $\mathcal{H}$ lie in the moment polytope $\Phi_T(M) \subset \mathbf{t}^*$. These will in general appear with some multiplicities m_β. We decompose the holomorphic sections of the prequantum line bundle as a direct sum of weights with multiplicities. In other words, we have*

$$\mathcal{H} = \oplus_{\beta \in \Lambda^W} m_\beta \mathbb{C}_\beta$$

for $m_\beta \in \mathbb{Z}^+$; in other words, the direct sum of m_β copies of the representation $\mathbb{C}_\beta$, which is a copy of $\mathbb{C}$ acted on by the representation β.

Proof This result is given by the Kostant multiplicity formula [4] and its generalizations due to Guillemin, Lerman and Sternberg [5]. As described in this book, we have

1. For toric manifolds, a weight appears with multiplicity 1 if and only if it is in $\Phi(M)$ (and 0 otherwise).
2. The multiplicity function $m : \Lambda^W \to \mathbb{Z}^{\geq 0}$ is related to the pushforward $\Phi_* \frac{\omega^n}{n!}$. The pushforward is obtained from the asymptotics of the multiplicity function under replacing ω by $k\omega, k \in \mathbb{Z}^+$. This operation dilates the moment polytope by k.   □

11.4 Holomorphic Bundles over G/T: The Bott–Borel–Weil Theorem

Theorem 11.19 (Kostant [4]) *Suppose $\lambda \in \mathbf{t}^*$. The symplectic form ω on the coadjoint orbit O_λ is integral if and only if $\lambda \in \Lambda^W \subset \mathbf{t}^*$.*

The Kirillov–Kostant–Souriau (or KKS) symplectic structure is

$$\omega([\lambda, X], [\lambda, Y]) = (\lambda, [X, Y])$$

where $X, Y \in \mathbf{g}$ and $\lambda \in \mathbf{t}$. (All elements of the tangent space at λ are of the form $[\lambda, X]$.) The group G acts transitively on the orbit, so we may assume $\lambda \in \mathbf{t}$ without loss of generality.

Let J be the almost complex structure on the orbit.

On a chart identified with a subset of the direct sum of the root spaces, each root space is identified with a copy of $\mathbb{C}$ and J acts by multiplication by i. We need to check that

$$\omega(J[\lambda, X], J[\lambda, Y])) = \omega([\lambda, X], [\lambda, Y]).$$

This is obvious because after this identification, J is simply multiplied by i.

Let $\lambda \in \Lambda^W$ be a weight for which $\mathrm{Stab}(\lambda) = T$. We may define a complex line bundle L_λ over $G/T \cong O_\lambda$ as follows. Define

$$\rho_\lambda \in \mathrm{Hom}(T, U(1))$$

by

$$\rho_\lambda(\exp(X)) = \exp(\lambda(X))$$

for $X \in \mathbf{t}$ and λ as above. We define

$$L_\lambda = G \times_{T, \rho_\lambda} \mathbb{C}$$

$$= (G \times \mathbb{C})/\sim \text{ where}$$

$$(g, z) \sim (gt^{-1}, \rho_\lambda(t)z)$$

for $t \in T$. Sections of L_λ are given by T-equivariant maps $f : G \to \mathbb{C}$; in other words, maps of the form

$$\{f : G \to \mathbb{C} | f(gt^{-1}) = \rho_\lambda(t)f(g)\}.$$

The action of G on the space of sections is

$$(g \cdot f)(hT) = f(ghT).$$

Theorem 11.20 *We have the following identification of homogeneous spaces: $G/T = G^{\mathbb{C}}/B$. Here $G^{\mathbb{C}}$ is the complexification of G (whose Lie algebra is the tensor product $\mathbf{g} \otimes \mathbb{C}$) and B (the Borel subgroup) is a complex Lie group defined by*

$$\mathrm{Lie}(B) = \left(\mathrm{Lie}(T) \otimes \mathbb{C}\right) \oplus \bigoplus_{\gamma>0} \mathbb{C}\gamma,$$

in other words, the Lie algebra of B is the direct sum of the positive root spaces and the complexification of the maximal torus. Recall that $\mathrm{Lie}(G) \otimes \mathbb{C}$ decomposes under the adjoint action of T as

$$\left(\mathrm{Lie}(T) \otimes \mathbb{C}\right) \oplus \bigoplus_{\gamma>0} \mathbb{C}_\gamma \oplus \bigoplus_{\gamma>0} \mathbb{C}_{-\gamma}.$$

Proof See, for example, Berline–Getzler–Vergne [6].   □

Here are some examples of complexifications of Lie groups:

$$SU(n)^{\mathbb{C}} = SL(n, \mathbb{C})$$

$$U(1)^{\mathbb{C}} = \mathbb{C}^*$$

$$U(n)^{\mathbb{C}} = GL(n, \mathbb{C})$$

Correspondingly, here are some examples of Borel subgroups. If

$$G = U(n),$$

its complexification is
$$G^{\mathbb{C}} = GL(n, \mathbb{C}).$$

The corresponding Borel subgroup B is the set of upper triangular matrices in $GL(n, \mathbb{C})$ (in other words, the set of $n \times n$ matrices with $z_{ij} = 0$ if $i > j$).

The groups $G^{\mathbb{C}}$ and B have obvious complex structures. So, therefore, does $G^{\mathbb{C}}/B$. This holomorphic structure is compatible with ω_λ (it comes from the almost complex structure J on $\mathrm{Lie}(G) \otimes \mathbb{C}$). The identity

$$\omega_\lambda([\lambda, X], [\lambda, Y]) = <\lambda, [X, Y]>$$

gives $\omega_\lambda(JZ_1, JZ_2) = \omega_\lambda(Z_1, Z_2)$. Here, the almost complex structure J is defined on $T_\lambda(G/T)$ and is defined at $T_{g \cdot \lambda}(G/T)$ by identifying this with

$$T_\lambda(G/T) \cong \oplus_{\gamma > 0} \mathbb{C}_\gamma.$$

This almost complex structure is integrable (in other words, it comes from a structure of complex manifold on G/T). Thus, L_λ acquires the structure of a holomorphic line bundle.

Lemma 11.21 *There is a homomorphism $p : B \to T_{\mathbb{C}}$.*

Proof The group B has a normal subgroup $N_{\mathbb{C}}$ for which $T_{\mathbb{C}} = B/N_{\mathbb{C}}$.   □

In the case $G = U(n)$, the subgroup $N_{\mathbb{C}}$ consists of the upper triangular matrices for which all entries on the diagonal take the value 1. In this case, the group $T_{\mathbb{C}}$ is the set of invertible diagonal matrices and the map p is projection on the diagonal.

Hence, $\rho_\lambda = \exp(\lambda) : T \to U(1)$ extends to $\rho_\lambda : T_{\mathbb{C}} \to \mathbb{C}^*$ and to $\bar{\rho}_\lambda : B \to \mathbb{C}^*$ via $\bar{\rho}_\lambda = \rho_\lambda \circ p$. Thus, we can define

$$L_\lambda = G_{\mathbb{C}} \times_{B, \rho} \mathbb{C}$$

$$= \{(g,z)\}/\sim$$

where $(g,z) \sim (gb^{-1}, \rho_\lambda(b)z)$ for all $b \in B$.

The space of holomorphic sections of L_λ is

$$H^0(O_\lambda, L_\lambda) = \{f : G^{\mathbb{C}} \to \mathbb{C} : f \text{ is holomorphic}, f(gb^{-1}) = \rho_\lambda(b)f(g)\}$$

for all $g \in G^{\mathbb{C}}$ and $b \in B$.

Theorem 11.22 (Borel–Weil–Bott [7, 8]) *If $\lambda \in \Lambda^W$ is in the positive Weyl chamber, then $H^0(O_\lambda, L_\lambda)$ is the irreducible representation of G with highest weight λ.*

11.5 Representations of $SU(2)$

We recall from (11.1) that the representations of $SU(2)$ arise by quantizing S^2:

$$H^0(M, L) = \{a_0 z_0 + a_1 z_1\}$$

$$H^0(M, L^k) = \{\sum_{j=0}^{k} a_j z_0^j z_1^{k-j}\}$$

The element

$$\tau := \mathrm{diag}(t, t^{-1}) \in SU(2)$$

acts on $\mathbb{C}^2$ by sending

$$\tau : \begin{pmatrix} z_0 \\ z_1 \end{pmatrix} \mapsto \begin{pmatrix} t z_0 \\ t^{-1} z_1 \end{pmatrix}$$

So $z_0^{k-j} z_1^{j} \mapsto t^{k-2j} z_0^{k-j} z_1^{j}$.

There are $k+1$ weights in total, each appearing with multiplicity 1.

References

1. J. Sniatycki, *Geometric Quantization and Quantum Mechanics* (Springer, 1980)
2. P. Gilkey, *Invariance Theory, the Heat Equation and the Atiyah–Singer Index Theorem* (Publish or Perish, 1996)
3. P. Griffiths, J. Harris, *Principles of Algebraic Geometry* (Wiley, 1994)
4. B. Kostant, Quantization and unitary representations, in *Lectures in Modern Analysis and Applications*, ed. by C. Taam. Lecture Notes in Mathematics, vol. 170 (Springer, Berlin-Heidelberg, New York, 1970), pp. 87–208
5. V. Guillemin, E. Lerman, S. Sternberg, *Symplectic Fibrations and Multiplicity Diagrams* (Cambridge University Press, 1996)

6. N. Berline, E. Getzler, M. Vergne, *Heat Kernels and Dirac Operators*. Grundlehren, vol. 298 (Springer, 1992)
7. R. Bott, Homogeneous vector bundles. Ann. Math. **66**, 203–248 (1957)
8. Encyclopedia of Mathematics. Bott–Borel–Weil Theorem. http://wnww.encyclopediaofmath. org/index.php?title=Bott-Borel-Weil_theorem&oldid=22175

Chapter 12
Flat Connections on 2-Manifolds

In this chapter, we aim to provide a survey on the subject of representations of fundamental groups of 2-manifolds, or in other guises flat connections on orientable 2-manifolds or moduli spaces parametrizing holomorphic vector bundles on Riemann surfaces.

The layout of this chapter is as follows. In Sect. 12.1, we describe background material. In Sect. 12.2, we provide a description of representations of the fundamental group of an orientable 2-manifold into the circle group $U(1)$. In Sect. 12.3, we describe the general case. Section. 12.4 describes Witten's formulas for the cohomology of these representation spaces. A sketch of the proof is given in Sect. 12.5. Sections 12.3 and 12.4 describe the topology of the spaces treated in Sects. 12.1 and 12.2, and may require more background in algebraic topology (for instance, familiarity with homology theory). Section 12.6 describes a particular class of Hamiltonian flows on spaces of representations. Finally, we describe geometric quantization of these spaces. The Verlinde formula gives the dimension of the quantization, and it will be outlined in Sect. 12.7.

12.1 Background Material

Let Σ be a compact two-dimensional orientable manifold. Unless otherwise specified, the dimension refers to the dimension as a *real* manifold. The space Σ can be described in different ways depending on how much structure we choose to specify.

1. Suppose first that we want only to specify the topological structure of the 2-manifold. These spaces are classified by their fundamental groups (in other words, by the genus r, for which the Euler characteristic of the space is $2 - 2r$):

© The Author(s), under exclusive licence to Springer Nature Switzerland AG 2019
S. Dwivedi et al., *Hamiltonian Group Actions and Equivariant Cohomology*,
SpringerBriefs in Mathematics,
https://doi.org/10.1007/978-3-030-27227-2_12

$$\pi = \pi_1(\Sigma^r) = \langle a_1, b_1, \ldots, a_r, b_r : \prod_{j=1}^{r} a_j b_j a_j^{-1} b_j^{-1} = 1 \rangle.$$

The a_j, b_j provide a basis of the first cohomology $H^1(\Sigma)$, chosen so that their intersection numbers are

$$a_j \cap b_j = 1$$

and all other intersections are zero.

2. Orientable surfaces may be endowed with structures of smooth orientable manifolds of dimension 2, and all smooth structures on a compact orientable 2-manifold of genus r are equivalent up to diffeomorphism.

3. Orientable 2-manifold may be endowed with additional structure, since they may be given a structure of complex manifold or Riemann surface (a Riemann surface is a complex manifold of complex dimension 1). It turns out that although there is only one class of smooth manifold corresponding to a surface of a given genus, there is a family of inequivalent ways of endowing such a surface with the structure of a complex manifold.

Correspondingly, there are three ways of describing the spaces we want to study, which turn out to be equivalent:

1. representations of the fundamental group into a Lie group, up to conjugation,
2. flat connections on Σ, up to the action of the gauge group, and
3. isomorphism classes of holomorphic bundles over Σ.

12.2   Cohomology of $U(1)$ Spaces

In this subsection, we describe the setup from the previous section in the simplest case, where the Lie group is the circle group $U(1)$.

Let $G = U(1)$. A connection A is simply a 1-form $\sum_{i=1}^{2} A_i dx^i$ on Σ. The holonomy of A around a cycle $\gamma(t)$ in Σ is

$$\exp\left(i \int_0^{2\pi} \sum_{j=1}^{2} A_j(\gamma(t)) \frac{d\gamma^j}{dt} dt \right),$$

since the parallel transport $x(\cdot)$ satisfies the equation

$$\frac{dx}{dt} = i A(\gamma(t)) x(t).$$

The solution to this equation is obtained by exponentiating the line integral of A along γ:

$$x(t) = \exp\left(i \int_{\gamma(0)}^{\gamma(t)} A(\frac{d\gamma}{dt'})dt'\right).$$

The connection A is flat if and only if $dA = 0$ in terms of the exterior differential d, which sends p-forms to $(p + 1)$-forms. The connection resulting from the action of an infinitesimal gauge transformation ϕ (where ϕ is a $\mathbb{R}$-valued function on Σ) is the 1-form $d\phi$.

The space of gauge equivalence classes of flat connections is isomorphic to $H^1(\Sigma, \mathbb{R})/\mathbb{Z}^{2r}$. In this case, the correspondence between flat connections and representations has an easy proof using Stokes' theorem: the parallel transport of a flat $U(1)$ connection around a closed loop is independent of deformations of the loop, so parallel transport gives a map from the space of flat $U(1)$ connections to the representations of the fundamental group.

We have

$$\mathcal{A}_{\text{flat}} / \exp \text{Lie}(\mathcal{G})$$

$$= \mathbb{R}^{2r}$$

Not all gauge transformations are in the image of the Lie algebra of the group of gauge transformations under the exponential map. To account for the action of those gauge transformations that are not in image of the Lie algebra, we must divide by an additional $\mathbb{Z}^{2r}$, which represents those gauge transformations which are not in the image of the Lie algebra under the exponential map.

$$\mathcal{A}_F/\mathcal{G} \cong \mathbb{R}^{2r}/\mathbb{Z}^{2r} \cong (U(1))^{2r}. \tag{12.1}$$

Thus, the cohomology of the space $U(1)^{2r}$ has $2r$ generators $d\theta_i$, $i = 1, \ldots, 2r$ and the only relations are that

$$d\theta_i \wedge d\theta_j = -d\theta_j \wedge d\theta_i, \quad i, j = 1, \ldots, 2r \tag{12.2}$$

12.3   Cohomology: The General Case

In this section, we describe the generators of the cohomology ring of representations of the fundamental group.

We describe the procedure to find analogous generators and relations for the case of M when G is a nonabelian group such as $U(n)$ or $SU(n)$.

Here, the generators of the cohomology ring are obtained as follows.

1. There is a vector bundle $\mathcal{U}$ (the "universal bundle") over $M \times \Sigma$. The bundle $\mathcal{U}$ has a structure of holomorphic bundle over $M \times \Sigma$, such that its restriction to $\{x\} \times \Sigma$ for any point $x \in M$ is the holomorphic vector bundle over Σ parametrized by the point x.

2. Take a connection A on $\mathcal{U}$ and decompose polynomials in its curvature F_A (for example, $\text{Trace}(F_A^n)$) into the product of closed forms on Σ and closed forms on $\mathcal{M}$.
3. Integrate these forms over cycles in Σ (a point or 0-cycle, the 1-cycles a_i and b_i, or the 2-cycle given by the entire 2-manifold Σ) to produce closed forms on $\mathcal{M}$, which represent the generators of the cohomology ring of $\mathcal{M}$.
4. These classes generate the cohomology of $\mathcal{M}$ under addition and multiplication.

For $G = SU(n)$, if n and d are relatively prime, define the space $M(n, d)$ as the space of representations of the fundamental group of Σ into $SU(n)$ (up to the action of G by conjugation) which send the product of commutators to $e^{2\pi i d/n}$ times the identity matrix.

One important cohomology class is the cohomology class of the symplectic form on $\mathcal{M}$. This cohomology class is often denoted by f and is the class obtained by taking the slant product of $c_2(\mathcal{U})$ with the fundamental class $[\Sigma]$ of Σ. Another important family of classes is that obtained by evaluating the classes on $\mathcal{M} \times \Sigma$ at a point in Σ.

12.3.1   The Case $M(2, 1)$

For the space $M(2, 1)$, the class obtained by evaluating $c_2(\mathcal{U})$ at a point in Σ is often denoted $a \in H^2(M(2, 1))$. This class is frequently chosen as the normalization of the cohomology class of the symplectic form on $M(2, 1)$. Newstead [1] describes the generators of this cohomology ring. The relations between these generators were established by Thaddeus [2].

12.4   Witten's Formulas

Witten [3, 4] discovered formulas for intersection numbers in the cohomology of these spaces $M(n, d)$. In particular, he obtained formulas for their symplectic volumes.

Let $G = SU(2)$ and set $n = 2$ and $d = 1$. In this case, these formulas are given by Donaldson [5] and Thaddeus [2]. The cohomology is generated by the generators $a \in H^4(M(2, 1))$, $f \in H^2(M(2, 1))$ and $b_j \in H^3(M(2, 1))$, $j = 1, \ldots, 2r$. The structure of the cohomology ring is then determined by the relations between these generators. Since the cohomology of a compact manifold satisfies Poincaré duality, these relations are determined by the intersection numbers of all monomials in the generators.

Donaldson and Thaddeus showed one may eliminate the odd degree generators b_j (see [2]), so the structure of the cohomology ring can be reduced to knowing the intersection numbers of all powers of the two even degree generators a and f described above.

$$\int_{M(2,1)} a^j \exp f$$

$$= \frac{(-1)^j}{2^{r-2} \pi^{2(r-1-j)}} \sum_{n=1}^{\infty} \frac{(-1)^{n+1}}{n^{2r-2-2j}} \qquad (12.3)$$

$$= \frac{(-1)^j}{2^{r-2} \pi^{2(r-1-j)}} (1 - 2^{2r-3-2j}) \zeta(2r - 2 - 2j).$$

Here we have used the notation

$$\exp f = \sum_{m \geq 0} \frac{f^m}{m!}$$

and we use the fact that $\int_{M(2,1)} \alpha = 0$ (where the integral denotes evaluation on the fundamental class) unless the degree of α equals the dimension of the space of conjugacy classes of representations. The $\zeta(n)$ denotes the Riemann zeta function.

We note that the formulas for intersection numbers can be written in terms of a sum over irreducible representations of G: this is the form in which these formulas appeared in Witten's work.

Example 12.1 The symplectic volume of the space M of gauge equivalence classes of flat G connections is given by the "Witten zeta function":

$$\int_M \exp(f) \sim \sum_R \frac{1}{(\dim R)^{2r-2}} \qquad (12.4)$$

where we sum over irreducible representations R of G. In the preceding formula and the next two formulas, the symbol $\sim$ means that the left-hand side is proportional to the right-hand side by a known proportionality constant. For the details, see [4]. In the special case of $SU(2)$, we have

$$\int_M \exp(f) \sim \sum_n \frac{1}{n^{2r-2}} \qquad (12.5)$$

and

$$\int_{M(2,1)} \exp(f) \sim \sum_n \frac{(-1)^{n+1}}{n^{2r-2}} \qquad (12.6)$$

where we sum over the irreducible representations of $SU(2)$, which are parametrized by their dimensions n. In this case, the Witten zeta function reduces to the Riemann zeta function.

Witten [3] expressed the symplectic volume of the moduli space in terms of Reidemeister–Ray–Singer torsion and gave a mathematically rigorous argument calculating it.

12.5   Mathematical Proof of Witten's Formulas

In this section, we give a very brief outline of the ingredients in the proof of Witten's formulas [6] for intersection numbers for the spaces $M(n, d)$.

The space M is a symplectic quotient

$$\mu^{-1}(0)/\mathcal{G},$$

where μ is the moment map (a collection of Hamiltonian functions whose Hamiltonian flows generate the action of a group $\mathcal{G}$ on a symplectic manifold M):

The space M may be constructed as an infinite-dimensional symplectic quotient of the space of all connections $\mathcal{A}$ by the gauge group $\mathcal{G}$: the moment map $\mu : \mathcal{A} \to \text{Lie}(\mathcal{G})^*$ of a connection A is its curvature $\mu(A) = F_A \in \Omega^2(\Sigma, \mathbf{g}) = \Omega^0(\Sigma, \mathbf{g})^*$ (and $\Omega^0(\Sigma, \mathbf{g})$ is the Lie algebra of the gauge group). Hence, the space M may be identified with $\mu^{-1}(0)/\mathcal{G}$.

The space M may also be constructed as a finite-dimensional symplectic quotient of a (finite-dimensional) space of flat connections on a punctured Riemann surface, by the action of the finite-dimensional group G. This may involve an extended moduli space [7] or the quotient of a space with a group-valued moment map [8].

We use formulas [6] for intersection numbers in a symplectic quotient; in terms of the restriction to the fixed points of the action of a maximal commutative subgroup of G (for $G = U(n)$ this subgroup is the diagonal matrices $U(1)^n$). The answer is given in terms of

1. the action of the maximal torus T on the normal bundle to the fixed point set to the T action;
2. the values of the moment map on the fixed point set;
3. the restriction of the cohomology classes to the fixed point set.

Using these methods, we recover Witten's formulas.

12.6   Hamiltonian Flows on the Space of Flat Connections on 2-Manifolds

In this section, we describe a collection of Hamiltonian flows on the spaces of conjugacy classes of representations introduced above. These were found by Goldman [9] and adapted by Jeffrey and Weitsman [10] to give Hamiltonian torus actions on an open dense subset.

A good reference for this material is [9].

Let $S_1, \ldots, S_{3g-3}$ be a collection of simple closed curves in a 2-manifold Σ of genus g. Each curve induces a Hamiltonian flow on the moduli space. For any two disjoint curves, the corresponding flows commute.

If the Hamiltonian Trace(θ) is replaced by $\cos^{-1}(\text{Trace}(\theta)/2)$, then the flows become periodic with constant period (they become the Hamiltonian flows for a Hamiltonian S^1 action).

12.7 Geometric Quantization of the SU(2) Moduli Space

A good reference for this section is [10].

We refer to the description of geometric quantization in the introduction to Chap. 11. The choice of $3g - 3$ disjoint circles in a 2-manifold (in other words a pants decomposition) specifies a real polarization on $\mathcal{M}$.

The map sending a flat connection A to the angle θ_j for which the holonomy of A around C_j is conjugate to

$$\begin{bmatrix} e^{i\theta_j} & 0 \\ 0 & e^{-i\theta_j} \end{bmatrix}$$

is the moment map for a Hamiltonian circle action on an open dense subset of $\mathcal{A}$.

Goldman [9] studied the Hamiltonian flows of the functions $A \mapsto \text{Trace}(\text{Hol}_{C_j} A)$ and found that these functions Poisson commute provided the curves S_j are disjoint.

In terms of flat connections, the Hamiltonian flows are given as follows. Let A be a flat connection on Σ. We assume a simple closed curve $C \subset \Sigma$ is chosen.

Assume the holonomy of A around C is in a chosen maximal torus T (say the diagonal matrices in $SU(2)$). It is always possible to apply a gauge transformation to A so that this condition is satisfied.

Define $\Sigma' = \Sigma \setminus C$. This has two boundary components C_+ and C_-.

Define $e^{it}(A)$ to be the result of applying a gauge transformation over Σ' which is the identity on C_- and the diagonal matrix with eigenvalues $e^{\pm it}$ on C_+. The result is a flat connection on Σ (because its values on C_+ and C_- are equal), but it is not gauge equivalent to A (since the gauge transformation does not come from a gauge transformation on Σ). This defines an S^1 action on (an open dense set of) the space of gauge equivalence classes of flat connections on Σ.

The circle action is not well defined when the stabilizer of the holonomy of A around C is larger than T, since in that case there is no canonical way to choose a maximal torus containing this holonomy.

We equip the genus 2 surface with the following pants decomposition (Fig. 12.1):

The moment polytope for the Hamiltonian torus actions in the genus 2 case then a tetrahedron.

The holonomies around C_j are characterized by the inequalities

$$|\theta_i - \theta_j| \le \theta_k \le \theta_i + \theta_j$$

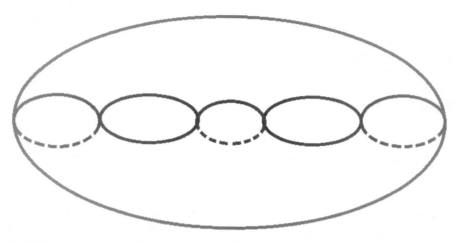

Fig. 12.1 Pants decomposition of a genus 2 surface

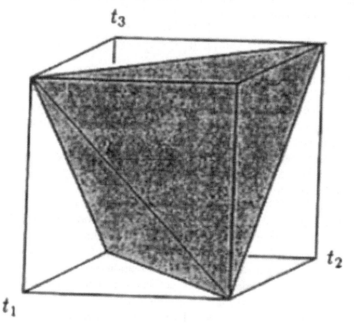

Fig. 12.2 The tetrahedron which is the image of the moment map in genus 2

$$\theta_i + \theta_j + \theta_k \le 2\pi$$

(where $0 \le \theta_i \le \pi$).

This set of inequalities specifies a tetrahedron (Fig. 12.2).

The Verlinde dimension is a formula for the space of holomorphic sections of $\mathcal{L}^k$ where $\mathcal{L}$ is the prequantum line bundle. Drezet–Narasimhan [11] showed that all line bundles are powers of the prequantum line bundle over $\mathcal{M}$.

The *Verlinde dimension formula* is a formula for the dimension of the space of holomorphic sections of $\mathcal{L}^r$ where $\mathcal{L}$ is the prequantum line bundle over $\mathcal{M}$. It is the number of labellings l_j of the curves C_j by integers in $[0, r]$ so that $(\frac{\pi l_i}{r}, \frac{\pi l_j}{r}, \frac{\pi l_k}{r})$ lies in the above tetrahedron whenever C_i, C_j, C_h are boundary circles of a pair of pants, and additionally

$$l_i + l_j + l_k \in 2\mathbb{Z}$$

The Verlinde formula is stated, for example, in Bismut–Labourie [12] and Witten [3]. These papers also give detailed references for the researchers who proved this formula.

In [10], the authors proved that the Verlinde dimension is the sum of integer values of the moment map for the circle actions.

Recall from toric geometry that the dimension of the space of holomorphic sections of $\mathcal{L}^k$ is the number of integer points in the dilation of the moment polytope (or Newton polytope) where the dilation factor is k. See for example [13].

This is an example of independence of polarization (since the dimension of the space of holomorphic sections of a line bundle is computed using a complex polarization, whereas the number of integer points in the moment polytope computes the dimension using a real polarization).

To get the correct answer, the authors of [10] needed to include points on the boundary of the moment polytope, where (strictly speaking) the Hamiltonian torus actions are not defined.

References

1. P. Newstead, Characteristic classes of stable bundles over an algebraic curve. Trans. Amer. Math. Soc. **169**, 337–345 (1972)
2. M. Thaddeus, Conformal field theory and the cohomology of the moduli space of stable bundles. J. Differ. Geom. **35**, 131–149 (1992)
3. E. Witten, On quantum gauge theories in two dimensions. Commun. Math. Phys. **141**, 153–209 (1991)
4. E. Witten, Two dimensional gauge theories revisited. J. Geom. Phys. **9**, 303–368 (1992)
5. S.K. Donaldson, Gluing techniques in the cohomology of moduli spaces, in *Topological Methods in Modern Mathematics. Proceedings of 1991 Conference in Stony Brook, NY in Honour of the Sixtieth Birthday of J. Milnor* (Publish or Perish, 1993)
6. L.C. Jeffrey, F.C. Kirwan, Intersection theory on moduli spaces of holomorphic bundles of arbitrary rank on a Riemann surface. Ann. Math. **148**, 109–196 (1998)
7. L.C. Jeffrey, Extended moduli spaces of flat connections on Riemann surfaces. Math. Ann. **298**, 667–692 (1994)
8. A. Alekseev, A. Malkin, E. Meinrenken, Lie group valued moment maps. J. Differ. Geom. **48**, 445–495 (1998)
9. W. Goldman, Invariant functions on Lie groups and Hamiltonian flows of surface group representations. Invent. Math. **85**, 263–302 (1986)

10. L.C. Jeffrey, J. Weitsman, Bohr–Sommerfeld orbits in the moduli space of flat connections and the Verlinde dimension formula. Commun. Math. Phys. **150**, 593–630 (1992)
11. J.-M. Drezet, M.S. Narasimhan, Groupe de Picard des variétés de modules de fibrés semi-stables sur les courbes algébriques. Invent. Math. 53–94 (1989)
12. J.-M. Bismut, F. Labourie, Symplectic geometry and the Verlinde formulas. *Surveys in Differential Geometry: Differential Geometry Inspired by String Theory*. Surv. Differ. Geom., vol. 5 (Int. Press, Boston, MA, 1999), pp. 97–311
13. W. Fulton, *Introduction to Toric Varieties* (Princeton University Press, 1993)

Appendix
Lie Groups

In this appendix, we collect together some standard facts about Lie groups which are used throughout the book. They are mostly presented without proof, and we direct the reader to Appendix B of [1] for more details.

Definition A.1 A Lie group is a group G which is also a smooth manifold, for which multiplication $m : G \times G \to G$ and inversion $i : G \to G$ are smooth maps. The identity element is usually denoted e.

Here are some standard examples of Lie groups:

Example A.2 The unitary group $U(1)$ with multiplication $m(e^{i\sigma}, e^{i\tau}) = e^{i(\sigma+\tau)}$ and inversion $i(e^{i\sigma}) = e^{-i\sigma}$.

Example A.3 The general linear group $GL(n, \mathbb{R})$ with multiplication $m(A, B)_{ij} = \sum_r A_{ir} B_{rj}$ and inversion $i(A)_{ji} = \frac{(-1)^{i+j}}{\det A} \tilde{A}_{ij}$, where $\tilde{A}_{ij}$ is the determinant of the matrix obtained by striking out the ith row and jth column of A.

Example A.4 The complex general linear group $GL(n, \mathbb{C})$ is exactly the same as $GL(n, \mathbb{R})$ with $\mathbb{R}$ replaced by $\mathbb{C}$.

Example A.5 The real numbers $\mathbb{R}$ with multiplication $m(a, b) = a + b$ and inversion $i(a) = -a$.

Definition A.6 A *Lie subgroup* of G is a regular submanifold which is also a subgroup of G.

Lie subgroups are necessarily Lie groups, with their smooth structure as submanifolds of G. The multiplication and inversion maps are automatically smooth. Lie subgroups are necessarily closed (see [2], Theorem III.6.18). Here are some examples of Lie subgroups:

© The Author(s), under exclusive licence to Springer Nature Switzerland AG 2019
S. Dwivedi et al., *Hamiltonian Group Actions and Equivariant Cohomology*,
SpringerBriefs in Mathematics,
https://doi.org/10.1007/978-3-030-27227-2

Example A.7   1. The orthogonal group

$$O(n) = \{A \in GL(n, \mathbb{R}) : AA^T = 1\}$$

is a Lie subgroup of $GL(n, \mathbb{R})$.

2. The special orthogonal group $SO(n) = \{A \in O(n) : \det(A) = 1\}$ is a Lie subgroup of $GL(n, \mathbb{R})$.

3. The unitary group $U(n) = \{A \in GL(n, \mathbb{C}) : AA^\dagger = 1\}$ is a Lie subgroup of $GL(n, \mathbb{C})$. Here $A^\dagger$ is the conjugate of the transpose of A.

4. The special unitary group $SU(n) = \{A \in U(n) : \det A = 1\}$ is a Lie subgroup of $GL(n, \mathbb{C})$.

Example A.8 Another example of a Lie group is the compact symplectic group, $Sp(n)$. The group is defined by

$$Sp(n) = \left\{ M(A, B) := \begin{bmatrix} A & -\bar{B} \\ B & \bar{A} \end{bmatrix} \right\}$$

where $A, B \in \text{End}(\mathbb{C}^n)$ and we insist that $M(A, B) \in U(2n)$. Equivalently

$$Sp(n) = \{U \in SU(2n) : \bar{U} J = J U\}$$

where $J = \begin{bmatrix} 0 & 1_n \\ -1_n & 0 \end{bmatrix}$.

The above examples give the complete list of the **classical Lie groups**. Namely,

- A_n ... $SU(n+1), n \geq 1$
- B_n ... $SO(2n+1), n \geq 2$
- C_n ... $Sp(n), n \geq 3$
- D_n ... $SO(2n), n \geq 4$.

The reason for the restriction on n is to avoid duplication: for low values of n many of the groups are isomorphic, or at least their Lie algebras are. For example, $SO(3)$ has the same Lie algebra as $SU(2)$. The classical Lie groups and a finite list of "exceptional Lie groups", namely, G_2, F_4, E_6, E_7, E_8, are the basic building blocks for compact connected Lie groups.

We now look at some more facts about Lie groups.

Theorem A.9 *If G_1 and G_2 are Lie groups and $F : G_1 \to G_2$ is a smooth map which is also a homomorphism, then* $\text{Ker}(F)$ *is a closed regular submanifold which is a Lie group of dimension* $\dim(G_1) - \text{rk}(F)$.

Proof This is Theorem III.6.14 of [2]. $\square$

Example A.10 $SL(n, \mathbb{R})$ is the kernel of $\det : GL(n, \mathbb{R}) \to \mathbb{R} \setminus \{0\}$.

Definition A.11 A Lie subgroup H of a Lie group G is a subgroup (algebraically) which is a submanifold and is a Lie group (with its smooth structure as an immersed submanifold).

Recall that $X \subset M$ is a regular submanifold if and only if there is a chart $\phi : U \to \mathbb{R}^m$ for which $\phi(U \cap X) = \phi(U) \cap \mathbb{R}^n$.

Proposition A.12 *A Lie subgroup that is a regular submanifold is closed. Conversely, a Lie subgroup that is closed is a regular submanifold.*

Definition A.13 Let F be a diffeomorphism and X a vector field on N, while Y is a vector field on M. Then X is F-related to Y if and only if $F_*(X_m) = Y_{F(m)}$ for all $m \in M$.

Proposition A.14 *The Lie brackets of F-related vector fields are F-related.*

Proof Suppose $F : N \to M$ is a diffeomorphism and X is a vector field on N, while Y is a vector field on M. Suppose that X_i, Y_i are F-related, meaning that $F_*(X_i) = Y_i$. We want to show that

$$F_*([X_1, X_2]) = [Y_1, Y_2].$$

For all $g \in C^\infty(M)$, and $x \in N$

$$(Y_i g)(F(x)) = (F_*)_x (X_i)(g) = X_i(g \circ F) \tag{A.1}$$

That is,

$$(Y_i g) \circ F = X_i(g \circ F).$$

Now let $f \in C^\infty(M)$ be arbitrary. Subbing into (A.1) Y_1 for Y_i and $Y_2 f$ for g, we obtain

$$Y_1(Y_2 f) \circ F = X_1((Y_2 f) \circ F).$$

Now apply (A.1) for $g = f$, $Y_i = Y_2$. This gives

$$Y_1(Y_2 f) \circ F = X_1(X_2(f \circ F)).$$

Likewise

$$Y_2(Y_1 f) \circ F = X_2(X_1(f \circ F)).$$

Hence,

$$([Y_1, Y_2]f) \circ F = [X_1, X_2](f \circ F)$$

so $[Y_1, Y_2]$ is F-related to $[X_1, X_2]$. $\qquad\square$

Recall that for $g \in G$ the map $L_g : G \to G$, defined by $L_g(h) = g \circ h$, is called the *left multiplication map*. For $Y \in T_e G$ define a vector field $\tilde{Y}$ by $\tilde{Y}_g = (L_g)_* Y$. By definition, we get

Proposition A.15 *The vector field $\tilde{Y}$ is smooth and left invariant.*

The vector field $\tilde{Y}$ is called the *left-invariant vector field* corresponding to $Y \in T_eG$.

Proposition A.16 *The bracket $[\tilde{X}, \tilde{Y}]$ is left invariant.*

Proof For any $h \in G$, the vector field $\tilde{Y}$ is L_h-related to itself by Proposition A.15. By the naturality of the Lie bracket, it follows that $[\tilde{Y}_1, \tilde{Y}_2]$ is also L_h-related to itself, for any $Y_1, Y_2 \in T_eG$. $\qquad\qquad\qquad\qquad\qquad\qquad\qquad\qquad\qquad\qquad\qquad\qquad\qquad\qquad\square$

We have shown that $[\tilde{Y}_1, \tilde{Y}_2] = \tilde{Z}$ for some $Z \in T_eG$. Hence, there is a *Lie bracket* operation $[\cdot, \cdot]$ on T_eG. The vector space T_eG equipped with $[\cdot, \cdot]$ is called the Lie algebra of G, denoted $\mathrm{Lie}(G)$ or $\mathbf{g}$.

Proposition A.17 *The tangent bundle TG of a Lie group G is trivial.*

Proof We have a global basis of sections given by the left-invariant vector fields. $\qquad\qquad\qquad\qquad\qquad\qquad\qquad\qquad\qquad\qquad\qquad\qquad\qquad\qquad\qquad\qquad\quad\square$

Example A.18 The tangent bundle TS^3 is trivial, since $S^3 = SU(2)$.

Theorem A.19 *For all $X \in T_eG$, there is a unique smooth homomorphism $\phi : \mathbb{R} \to G$ with $\frac{d\phi}{dt}\big|_{t=0} = X$.*

Proof Given X, we construct the corresponding left-invariant vector field $\tilde{X}$. Take the integral curve $\phi : (-\epsilon, \epsilon) \to G$ through e, with $\phi(0) = e$. Extend it to $\phi : \mathbb{R} \to G$ by defining

$$\phi(t) = \phi(\epsilon/2) \circ \ldots \phi(\epsilon/2)\phi(r)$$

where there are k copies of $\phi(\epsilon/2)$, and $t = k(\epsilon/2) + r$. Then $t \mapsto \phi(s) \cdot \phi(t)$ is an integral curve of $\tilde{X}$ passing through $\phi(s)$ at $t = 0$. We also have that $\phi(s + t)$ is such an integral curve. Therefore, by uniqueness of integral curves

$$\phi(s + t) = \phi(s) \cdot \phi(t).$$

Conversely if $\phi : \mathbb{R} \to G$ is a smooth homomorphism, and $f : G \to \mathbb{R}$ is smooth, then $d\phi/dt$ is a tangent vector to G at $\phi(t)$. Recall

$$
\begin{aligned}
\frac{d\phi}{dt}(f) &= \lim_{h\to 0} \frac{f(\phi(t+h)) - f(\phi(t))}{h} \\
&= \lim_{h\to 0} \frac{f(\phi(t)\phi(h)) - f(\phi(t))}{h} \\
&= \frac{d}{du}\Big|_{u=0} f \circ L_{\phi(t)} \circ \phi(u) \\
&= (L_{\phi(t)})_* \frac{d}{du}\Big|_{u=0} (f \circ \phi(u))
\end{aligned}
$$

$$= (L_{\phi(t)})_* X(f)$$
$$= \tilde{X}(\phi(t))(f).$$

Thus ϕ is an integral curve of $\tilde{X}$.   □

Definition A.20 A one-parameter subgroup of G is a homomorphism $\phi : \mathbb{R} \to G$.

We have thus shown that there is a bijective correspondence between Left-invariant vector fields and one-parameter subgroups.

Given $X \in \mathbf{g}$, let ϕ be the unique smooth homomorphism with $\frac{d\phi}{dt}(0) = X$. Then we define the exponential map as follows.

Definition A.21 With the above notation, we define $\exp : \mathbf{g} \to G$ by

$$\exp(X) = \phi(1).$$

The map exp is called the exponential map.

Clearly

$$\exp(t_1 + t_2)X = (\exp t_1 X)(\exp t_2 X)$$

and

$$\exp(-tX) = (\exp tX)^{-1}.$$

Proposition A.22 *The map* $\exp : \mathbf{g} \to G$ *is smooth, and 0 is a regular value. Thus,* exp *takes a neighbourhood of* $0 \in \mathbf{g}$ *diffeomorphically onto a neighbourhood of* $e \in G$.

Proof Note that $T_{(X,a)}(\mathbf{g} \times G) \cong T_e G \oplus T_a G$. Define a vector field Y on $\mathbf{g} \times G$ by

$$Y_{(X,a)} = 0 \oplus \tilde{X}(a).$$

Then Y has a flow

$$\alpha : \mathbb{R} \times (T_e G \times G) \to T_e G \times G$$

which is smooth, since Y is smooth. Since $\exp(X)$ is the projection on G of $\alpha(1, 0 \oplus X)$, exp is smooth as it is the composition of smooth maps.

Given $v \in T_e G$, the curve $c(t) = tv$ in $T_e G$ has tangent vector v at 0. So

$$\exp_0(v) = \frac{d}{dt}\bigg|_0 \exp(tv) = v.$$

Hence

$$(d \exp)|_0 = \text{id}.$$

So exp is a diffeomorphism in a neighbourhood of 0.

Proposition A.23 *If $\psi : G \to H$ is a homomorphism, then*

$$\exp_H \circ \psi_* = \psi \circ \exp_G.$$

Proof If $\psi : G \to H$, and $X \in T_e G$, then let $\phi : \mathbb{R} \to G$ be a homomorphism with

$$\left. \frac{d\phi}{dt} \right|_{t=0} = X.$$

Then $\psi \circ \phi : \mathbb{R} \to H$ is a homomorphism with

$$\left. \frac{d}{dt}(\psi \circ \phi) \right|_{t=0} = \psi_* X.$$

Hence

$$\exp(\psi(X)) = \psi \circ \phi(1) = \psi(\exp X),$$

as desired.

Proposition A.24 *If $G = GL(n, \mathbb{R})$ then $\text{Lie}(G) = M_{n \times n}(\mathbb{R})$, the vector space of $n \times n$ real matrices, and*

$$\exp(X) = \sum_{n \geq 0} \frac{X^n}{n!}. \tag{A.2}$$

Proof We define a norm on $\text{Lie}(G)$ by

$$|X| = \sup_{1 \leq i, j \leq n} |x_{ij}|.$$

Since $|AB| \leq n|A||B|$, it follows that

$$|X^k| \leq \frac{1}{n}(n|X|)^k.$$

Hence, the series (A.2) converges absolutely. Also, the one-parameter subgroup of $GL(n, \mathbb{R})$ whose left-invariant vector field has the value X at e is $\exp(tX)$ since

$$\sum_{n \geq 0} \frac{t^n X^n}{n!} = \text{id} + tX + O(t^2),$$

showing that

$$\frac{d}{dt}\bigg|_{t=0} \sum_{n \geq 0} \frac{t^n X^n}{n!} = X.$$

□

Proposition A.25 *If $G = GL(n, \mathbb{R})$ and $A, B \in \text{Lie}(G)$ then*

$$[A, B] = AB - BA.$$

Proof Arbitrary elements $A, B \in \text{Lie}(G)$ are of the form

$$A = \sum_{i,j} a_{ij} \frac{\partial}{\partial x_{ij}},$$

$$B = \sum_{i,j} b_{ij} \frac{\partial}{\partial x_{ij}}$$

where a_{ij}, b_{ij} are constants. Let $\tilde{A}, \tilde{B}$ be the left-invariant vector fields corresponding to A and B. Then, by definition of the Lie bracket on vector fields,

$$[\tilde{A}, \tilde{B}]f = \tilde{A}(\tilde{B}f) - \tilde{B}(\tilde{A}f).$$

If $x \in GL(n, \mathbb{R})$, then

$$\tilde{B}(x)_{ij} = (xB)_{ij} = \sum_r x_{ir} b_{rj}$$

so

$$A(\tilde{B}f) = \sum_{i,j} \sum_{k,\ell} a_{k\ell} \frac{\partial}{\partial x_{k\ell}}$$

$$= \sum_r a_{ir} b_{rj} \frac{\partial}{\partial x_{ij}} f \ + \ \text{terms with } \frac{\partial}{\partial x_{k\ell}} \frac{\partial}{\partial x_{ij}}$$

Likewise

$$B(\tilde{A}f) = \sum_r b_{ir} a_{rj} \frac{\partial}{\partial x_{ij}} f.$$

It follows that

$$[\tilde{A}, \tilde{B}] = \widetilde{AB - BA}.$$

□

Proposition A.26 *For matrix groups, if* $[X, Y] = 0$ *then* $\exp(X + Y) = \exp X \exp Y$.

Proof For matrix groups, using Eq. (A.2), we have that

$$\exp(X + Y) = \sum_{n \geq 0} \frac{(X + Y)^n}{n!}$$

$$= \sum_{m=0}^{\infty} \sum_{p=0}^{m} \frac{1}{(m - p)!} X^{m-p} \frac{1}{p!} Y^p$$

$$= \Big(\sum_{k=0}^{\infty} \frac{1}{k!} X^k \Big) \Big(\sum_{\ell=0}^{\infty} \frac{1}{\ell!} X^\ell \Big)$$

$$= \exp X \exp Y.$$

$\square$

References

1. V. Guillemin, V. Ginzburg, Y. Karshon, *Moment Maps, Cobordisms and Hamiltonian Group Actions*. Mathematical Surveys and Monographs, vol. 98 (AMS, 2002)
2. W. Boothby, *An Introduction to Differentiable Manifolds and Riemannian Geometry*. Pure and Applied Mathematics, vol. 120 (Academic Press, 1986)

Index

© The Author(s), under exclusive licence to Springer Nature Switzerland AG 2019
S. Dwivedi et al., *Hamiltonian Group Actions and Equivariant Cohomology*,
SpringerBriefs in Mathematics,
https://doi.org/10.1007/978-3-030-27227-2

Printed in the United States
By Bookmasters